The Titan Guide to:

Legal Issues
in a Digital World

A Practical Toolkit for
Content Creators and Influencers

Jeffrey L. Wilson, Esq.

First Edition

TITAN LAW PLLC

Book Cover Inspired by Mia Wilson

First Edition – January 2025

Printed in the United States of America.

ISBN: 979-8-218-60401-1

For more information, or to book an event, visit :

TITAN LAW PLLC

www.titanlawny.com

With profound appreciation to
Moli for her incalculable patience and belief
and to Mia and James
for their countless hours of inspiration and joy.

CONTENTS

Introduction

Welcome to the wild, wonderful world of content creation! Yes, I know, that's pretty lame. You should get used to it. There's going to be a lot more "lame" crap like that in the pages that follow! Plus, not for nothing, some pure gold. It will be your job to sift the one to get to the other. Let's do this.

Whether you're a YouTuber with a dream and a caffeine problem, a TikTok/Only Fans star in the making, or someone who's just really good at turning your emotional support donkey's antics into viral memes, this book is for you. Some of you may be asking yourself one question: why am I consuming information in book form? Is that still a thing? Technically, that's two questions and, yes, it's still a thing.

Let me tell you a bedtime story. Before there were doorbell cameras, before there were viral videos of plumbers and teenage boy fights, before Instagram destroyed the lives of countless young girls with equal parts envy and shame, and before the Chinese Communist Party figured out how to brainwash an entire generation of Americans with only a few strange cartoons and a video platform designed to celebrate dancing in one's underpants, there were things called books. People wrote them (they're called authors). People read them (they're called readers). And, some of those readers even learned a few things from those authors. For your generation, I'm sure it all sounds like ancient technology. I get it. Try it anyway.

In any event, congratulations on stepping into one of the most exciting (and occasionally terrifying) career paths out there. You couldn't have picked a more fickle, odd, delightful, low percentage, and soul-crushing business if you tried. Which is why I admire you. In the face of daunting odds, you want to charge forward anyway. That's brave (not dodging-bullets brave, but way better than sitting-on-your-mother's-couch-complaining brave). It takes energy and drive and ambition. So, good for you. Many of you have what it takes to make moderate to decent incomes while not working too hard. Some of you will have to go back to college or to the construction site and get real jobs. And, at least one of you is going to blow up into a *bona fide* superstar (seriously). Now, let's talk about not ruining it with a bunch of lawsuits and legal hassles and wasted money.

Here's the thing about being a creator: It's not just about the creative stuff. Sure, your amazing content will have people hitting "Subscribe" and smashing that "Like" button, but behind the scenes, there's a whole lot of unglamorous, unfiltered, and undeniably important legal stuff to figure out. And if you're not careful, you could find yourself tangled in a legal crap-storm faster than you can spell "copyright infringement."

Why You Need (to Love) This Book

Most creators don't start out thinking, "You know what's fun? Reading about contracts and copyright law." And yet, understanding the basics of intellectual property, contracts, and platform policies is absolutely essential if you want to survive (and thrive) as a creator. There are two ways to learn the lessons you're going to need to succeed in this business: get ripped off by people who already know the lessons or, alternatively, spending some time to find the landmines before you start tearing off through the field.

The good news? This book takes all that intimidating legal jargon and translates it into "regular" English you can understand. I'm just going to assume you understand "regular" English otherwise this will be a huge waste of time for both of us. The better news? I've packed it with real-world examples, actionable advice, and just enough humor to keep you from crying into your espresso martini.

You'll learn:

- How to protect your work from copycats, scammers, and that one guy who thinks he can "remix" your content without asking.

- Why contracts are your best friend (even if they look like they were written in a dead language from a different era).

- The difference between "monetizing your content" and "trading your hard work for coffee mugs and free sweatshirts."

- What to do when the internet inevitably tries to cancel you (because let's face it, it happens to everyone). Except you, Kim Kardashian. It can't seem to get rid of you no matter how hard it tries.

What This Book Is Not

This book is not a replacement for a lawyer. I know, I know, I'm a lawyer, so I have to say that. No, seriously, I have to say that.

While I'm here to educate, entertain, and give you some solid tips to navigate the murky waters of legal issues in the creator space, every situation is different. Think of this as a survival guide or toolkit. Do not think of this book as an all-access pass to ignoring professional legal help when you really need it.

Let's also be super clear here: **I'm a lawyer, but I'm not your lawyer**. Nothing in here should be considered legal advice and it should be used for educational purposes only. It contains a fair number of opinions – some of which you may or may not publicly agree with (though you know I'm right, privately). If you want, just consider the whole book to be a form of shameless promotion or roughly 200 pages of blowhard self-importance. #attorneyadvertising.

Who is this guy? Who does he think he is? How come I've never seen his Instagram feed? Just a few questions you might be asking. Fair points, all. Also @titanlawny, in case you were wondering.

As a measure of background, I've been a New York City-based commercial litigation attorney for almost 20 years. I run my own law firm (TITAN LAW PLLC), specializing in all sorts of disputes and areas of the law, including the ones faced by content creators, artists, athletes, and influencers. I also wear bow ties. From those 3 facts alone, you can rightly assume the following: 1) I'm good at yelling at people about small stuff; 2) I love contracts and writing nasty letters; and 3) I'm probably too old to be writing this book. Fear not. According to my wife, I'm often ridiculously immature, despite my advanced age and education. So, that probably makes me the perfect person to write this book.

I have represented every kind of business, from single owners to multi-billion dollar municipal entities, from small businesses to industry giants, from individuals to entire cities, and each one presents a different challenge. But, the rules are the same for everyone. That's good news for you. It means you're going to get the benefit of wisdom a lot of other people have already paid for. And all you had to shell out was the extremely reasonable price of this book.

Thousands of contracts, hundreds of lawsuits, dozens and dozens of clients – they've all taught me one immutable lesson about human nature: we are all mostly stupid. Some of us fake it better than others, but the lesson is the same. We don't like to admit when we are confused by something, and we try to pretend we know more than we do. On top of that, we're all fairly lazy. That's why I wrote this book. The law is excruciatingly complex, and, as you pull one layer of the onion away, another is right there waiting for it. This book could be an entire series of books, filled with statutes, case law, judicial decisions, and footnotes up the ying-yang.

Don't worry. This book isn't going to contain any of that. We're going to try to take some difficult and confusing concepts and distill them down to a few practical tips and tricks you can put to good use right away. If your interest is piqued, there are libraries full of more detail on each topic. A fairly normal contract law textbook in a fairly normal law school could be 1,000 pages long. What I gloss over in terms of legal detail in this book, I will attempt to make up with sarcasm, oversimplification, and good old-fashioned personal bias.

Also, if you're reading this book in the bookstore without paying for it, you suck.

A Few Ground Rules

Before we dive in, let's establish some ground rules:

1. **Don't Panic.** Legal stuff can feel overwhelming, but I promise it's manageable. Take it one step at a time.

2. **Ask for Help.** Whether it's a lawyer, an accountant, or your tech-savvy cousin, don't be afraid to call in reinforcements.

3. **Keep Creating.** At the end of the day, your creativity is your superpower. This book is here to make sure you can keep using it safely and successfully.

A Quick Note About Humor

Yes, this book is funny - or at least I think it is. Why? Because the only thing worse than dealing with legal headaches is dealing with them without a sense of humor. So buckle up, laugh a little, and let's make sure your content empire doesn't turn into a cautionary tale.

Ready? Alrighty then.

By the time you finish this book, you'll know how to navigate the legal side of content creation like a pro. You'll have the tools to protect your work, avoid common pitfalls, and build a brand that's as strong legally as it is creatively. You'll have a resource you can refer back to as you navigate along your journey. And most importantly, you'll have a lot fewer reasons to lose sleep over that ominous "notice of infringement" e-mail or "cease-and-desist" letter.

Let's get started. The internet awaits!

CHAPTER 1

Understanding the Basics of Intellectual Property (IP)

Because there are really no shallow ends in the legal issue pool, I'm going to just push you into the deep end. Not literally, of course. Never, ever push a woman into a pool unless you have health insurance, a foolproof exit strategy, and probably a good lawyer. I've got at least two of those, so let's go.

What is Intellectual Property (IP)?

Most people understand the concept of property – look around at all your stuff, the contents of your room. That's your tangible property (such as it is). Half-full cans of Red Bull and broken PS5 controllers count, too. It's tangible, meaning you can touch it. As a content creator, you have stuff you need to do your thing that fall in this bucket – your phone, laptop, lighting rig, microphones, bikini collection, etc. You take care of them because you need them. What about all that stuff in your brain, though? You can't touch that stuff, but it is the most important set of assets (i.e., property) you own. Even if all you do is make shuffle dance videos. That "intangible" asset is known as "**intellectual property**" (also called IP).

IP refers to creations of the mind that are given legal protections to safeguard the rights of their creators. These creations can take many forms, including artistic works like paintings, music, and literature; technological inventions and innovations; designs for products or processes; distinctive logos or symbols that represent a brand; and even the names of businesses or products. More specifically, maybe, these creations are the little nuggets of genius you film in skate parks, the front seat of your car, on public sidewalks, and in rented private jets you pretend you own. Anything your pot-addled brain can dream up between skipping class and binge-watching the Love Boat counts as IP. You probably call it "content." By granting these protections, IP laws encourage creativity and innovation while ensuring that creators and businesses can benefit from their efforts.

IP can be distinctive and high-brow (think of the logo for THE NEW YORK TIMES) or hilarious and low-brow (think "git-R-done"). Larry the Cable Guy, in fact, has 23 federally registered trademarks! Whether or not you wear a worn-out camo baseball cap like Larry the Cable Guy, your IP is what makes your thing yours, different from what came before, knowing that it will be protected after you release it on the world. It's your brand, your calling card, your jam, and your particular stank.

As a content creator, your IP could include:

- Videos and scripts.

- Logos and branding materials.

- Photos, designs, and artwork.

- Courses or educational content.

- Even memes (yes, memes).

Think of IP as the legal shield for your creative work, giving you control over how it's used, shared, or monetized. Without IP protection, others could exploit your hard work without permission, potentially impacting your reputation and income.

BUT: It's not a "set it and forget it" type of thing (fun fact: Popeil Inventions, Inc., maker of the Ronco Rotisserie Oven, attempted to trademark that exact saying but ultimately abandoned its application). Lurking in the deep background is something called the **"public domain,"** which refers to creative works that are no longer protected by copyright laws, meaning they are free for anyone to use without needing permission or paying royalties. This can occur when a copyright expires after a set period, or when the creator voluntarily gives up their rights.

Works in the public domain can include books,

music, art, and even inventions. Because these works are no longer under copyright protection, they can be reproduced, modified, and distributed by anyone, which allows for widespread use and adaptation. However, it's important to note that public domain status can vary depending on the laws of a particular country and the specific type of work. But, it will take a while before your clothes-changing videos ever get there, so relax.

Think even a big company can protect its IP from falling into the public domain? Think again. Ask Disney. Disney is beginning to lose some copyright protection over its iconic characters, Mickey Mouse and Winnie the Pooh, due to the expiration of certain copyrights. Mickey Mouse, originally created in 1928, has started to enter the public domain in some of his earliest forms, meaning that the character's appearance in his original iteration no longer enjoys the full copyright protection it once did. As a result, others can now legally use versions of Mickey from his earliest films in creative works without violating Disney's copyrights. They can make him do all sorts of stuff (e.g., think slasher films). Similarly, Winnie the Pooh, originally created by A.A. Milne in 1926, has seen parts of the character's story and imagery enter the public domain, which allows others to use certain aspects of the honey-loving bear without Disney's permission. That said, Disney still retains exclusive rights to the more recent versions of both characters, including later films and adaptations, giving them continued control over their modern depictions.

Don't even get me started on the "Happy Birthday" song. That did not hit the public domain until 2016 and, even now, there are still some IP protections. The lyrics? Public domain, so you don't have to worry about getting a license for grandma's birthday card. The musical arrangement? That depends. Outside of the U.S.? Also not crystal clear.

Now you know why lawyers continue to exist.

The Four Types of IP

Here's a detailed breakdown of the key types of IP and how they relate to your content:

1. **Copyright.** Protects original works of authorship, like videos, blogs, music, or artwork. Copyright is automatic when you create something, but registration of your copyright provides even stronger protections.

   FYI, there is no magic around the © symbol. Any time you create something, it's copyrighted. If you put the © symbol on it, great. That's good practice. If not, you are still protected (remember, copyright protection is "automatic."). If you really want the most protection possible, you will need to register the copyright (more on that below).

- *Practical Tip:* If you're creating a series of videos or a recurring podcast, consider registering the entire series as a single work to save costs.

- *Additional Insight:* Copyrights last for the creator's lifetime plus 70 years in most countries, making it a long-term asset for your brand.

2. **Trademarks.** Protects brand identifiers like names, logos, and slogans. For example, Nike's "Just Do It" is trademarked. For creators, this could mean registering the name of your YouTube channel or the logo of your personal brand.

- *Practical Tip:* Before deciding on a name or logo, conduct a trademark search to ensure it's not already in use. Clearly, Meghan Markle didn't read this book before getting kicked out of the U.S. Patent & Trademark Office (USPTO) with her ideas for the American Riviera Orchard brand – a couple of times.

- *Additional Insight:* Trademarks can also include non-conventional elements like unique packaging (trade dress) or even sounds, such as the MGM lion's roar. I have no comment with regard to Meghan Markle, here.

- *Impress Your Friends:* If you see a ® symbol, that means the USPTO has registered a trademark and the applicant has exclusive rights to use it (and to sue you for using it). Do not just put that symbol on your logo without registering it and expect it to mean anything. Similarly, the ™ symbol means that someone has claimed that mark as a trademark (i.e., by filing an application with USPTO), but does not yet have a registration.

3. **Patents.** Protect inventions or processes, like a unique way of editing videos or creating a new app. While patents may not apply to all creators, they're essential for those developing innovative technology or tools. This is a specialty area of the law, and you should consult a patent attorney if you are thinking about getting one. Or, just call me; I'll give you the name of a good one.

  - *Example:* If you create a software plugin that automates video editing, you may want to explore patent protection.

- *Additional Insight:* Patents require detailed applications and can take years (and $$$) to secure, but they provide robust protections for unique inventions.

4. **Trade Secrets.** Protects confidential business information, like a proprietary recipe or a unique way you monetize your content. Unlike other forms of IP (which require disclosure by their very nature), trade secrets aren't registered—they're protected by keeping them confidential. If you tell everybody your secret sauce for camera tricks that skirt the Instagram content rules, well it's not "secret" anymore, is it?

   - *Example:* A social media creator's proprietary strategy for going viral could be considered a trade secret. Incidentally, I would be willing to exchange some free legal work for the ability to license that trade secret.

   - *Additional Insight:* Trade secrets are only valuable as long as they remain secret, so sign confidentiality agreements with anyone who has access. I have included a simple non-disclosure agreement at the end of the book which you can use for free. You're welcome.

Why IP Matters for Creators

Understanding and protecting IP is critical for:

1. **Monetization.** Without IP protection, it's easy for others to copy and profit from your work. For example, if you design a popular logo and fail to trademark it, someone else could sell merchandise with your design and keep the profits.

   - *Example:* Imagine creating a catchy slogan for your podcast that gains traction. Without trademarking it, another entity could use the slogan for their own merchandise line, capitalizing on your brand's popularity.

   - *Detailed Example:* MrBeast® (note the wonky spacing and lack of punctuation – sometimes, that's going to make all of the difference at the USPTO). Burgers, philanthropy, television shows, etc. That guy is an object lesson in monetizing protected IP.

2. **Preventing Infringement.** It's not just about protecting your own work; understanding IP helps you avoid accidentally violating others' rights. For example, using copyrighted music in

your videos without permission could result in takedowns or legal action.

- *Example:* A content creator borrowed a 10-second clip of a song for a tutorial video. Even though it was a short segment, it triggered an automated copyright claim, demonetizing the video.

- *Detailed Example:* In one of the most famous recent cases, Robin Thicke and Pharrell Williams were sued by the family of Marvin Gaye, who argued that their song "Blurred Lines" copied elements of Gaye's 1977 song "Got to Give It Up." The court sided with Gaye's family and awarded them $7.4 million in damages, later reduced to $5.3 million.

3. **Building Your Brand.** Strong IP protections can make your brand more valuable. Trademarks, in particular, help distinguish your content and prevent others from diluting your brand's identity.

- *Example:* A creator's unique logo became synonymous with their tutorials. Trademarking the logo not only prevented imitators but also increased its licensing potential.

- *Detailed Example:* TED Talk®. A strong brand is so much easier than saying, "TED (Technology, Entertainment, Design) is a nonprofit organization that hosts talks featuring influential speakers across various fields" every time you bring it up.

4. **Legal Security.** Having IP protections in place means you're better equipped to take action if someone misuses your content. Whether it's a copyright infringement or unauthorized use of your logo, legal recourse is much stronger when your IP is properly protected.

   - *Example:* An artist's digital illustrations were reposted on a merchandise website without permission. Because the works were registered, the artist successfully issued a cease-and-desist letter, claiming monetary damages.

   - *Detailed Example:* See above re: "Blurred Lines." You can also ask Kanye West, Led Zeppelin, Little Nas X, Katy Perry, The Clash, The Verve, and Sam Smith about wading into legal messes with content disputes. Well, you can ask their lawyers, anyway.

Common IP Myths for Creators

Let's dive deeper into some widespread misconceptions (the internet is full of them):

1. **"It's on the internet, so it's free to use."**

   - *Reality*: Most online content is protected by copyright, even if it's not explicitly labeled. For example, using a photo from a blog post without permission could land you in legal trouble, even if the blog doesn't have a copyright notice.

   - *Further Detail:* Public domain content is the exception, but always verify the source to ensure it's genuinely free to use. There are sites that make copyrightable images available so long as you properly attribute the source or the photographer. Some good examples include: Unsplash, Pexels, Pixabay, Burst, Reshot, and Canva. Other examples are available – do your own Google search.

2. **"I changed it, so it's not infringement."**

   - *Reality*: Copyright law covers "derivative works," which means that even altered versions of copyrighted material can still infringe. For instance, remixing a song without permission could lead to a copyright claim.

- *Further Detail:* Transformative use, such as parody or critique, might qualify as fair use, but it's a legal gray area that varies by jurisdiction. We'll talk more about "fair use" in Chapter 2.

3. **"If I'm not making money, I'm safe."**

- *Reality*: Infringement is about unauthorized use, not profit. For example, reposting a copyrighted image without permission is still a violation, even if you're not monetizing it.

- *Further Detail:* Non-commercial use might reduce damages but doesn't eliminate liability.

4. **"I can use anything if I give credit."**

- *Reality*: Giving credit doesn't replace the need for permission. For example, using a copyrighted video clip in your content without permission isn't excused by merely crediting the original creator.

- *Further Detail:* Some licenses, like Creative Commons, allow use with attribution, but always check the specific terms.

Examples of IP Disputes in the Creator World

To bring this to life, here are detailed real-world examples:

1. **The React Video Scandal.** A popular YouTube channel attempted to trademark the word "react" and monetize reactions. The backlash from creators and viewers highlighted the importance of not overstepping when filing for trademarks.

    *   *Lesson Learned:* Trademarks should be specific and not overly broad to avoid public backlash and rejection.

2. **Copyright Claims Gone Wild.** One content creator's original music was falsely flagged for copyright infringement by an automated system. The dispute took months to resolve, costing time and money. This emphasizes the need to understand and navigate copyright claim systems.

    *   *Lesson Learned:* Registering your work and using digital rights management tools can help assert ownership.

3. **The Instagram Artwork Debate.** Many artists found their work reposted without credit by large accounts. This sparked a movement for

creators to demand recognition and legal action against unauthorized use.

- *Lesson Learned:* Watermarking and monitoring tools can deter misuse and protect your work.

Next Steps: Building Your IP Toolkit

To protect your IP effectively:

1. **Register Copyrights.** While copyright is automatic, registering significant works provides stronger legal protections.

2. **Trademark Your Brand.** If you're building a recognizable brand, trademarking your name, logo, or slogan is crucial.

3. **Monitor for Infringement.** Use tools like reverse image searches to track unauthorized uses of your content.

4. **Seek Legal Help When Necessary.** Consult a lawyer for complex issues or disputes.

5. **Leverage Technology.** Platforms like YouTube's Content ID can automatically detect and address unauthorized use of your videos.

These are only the basics, but, hopefully, this introduction gives you a good sense of the landscape to enable you to be better equipped to protect your creations and navigate the digital landscape confidently. In the next chapter, we'll explore copyright in more detail, including practical tips for avoiding infringement and handling disputes.

CHAPTER 2

Copyright Essentials

Understanding Copyright

Copyright is one of the fundamental pillars of IP law, providing creators with exclusive rights over their original works (i.e., those great ideas you have). For content creators, this protection applies to:

- Written content (blogs, scripts, eBooks).

- Visual media (videos, photos, artwork).

- Audio content (music, podcasts, sound effects).

Once you put them in a tangible format (e.g., page, photo, video), by default, copyright grants the creator the *exclusive* right to reproduce, distribute, perform, display, and create derivative works based on their creation. It also entitles the creator the right to license those works to others (i.e., to monetize their existing IP rights). How long you retain those exclusive rights depends on a few factors:

- Credited works created after January 1, 1978 – protection lasts the life of the author plus 70 years (for joint works, 70 years after the death of the last author alive);

- Anonymous works – protection lasts for 95 years from publication or 120 years from creation (whichever is less);

- Credited works created before January 1, 1978 – call your lawyer. There's an elaborate interplay between various copyright laws passed in 1909 and 1976 that determines the ultimate protection. Depending on what you're doing with the copyrighted material (e.g., movie reviews, video montage creation), you could run into issues here.

As you may have surmised, copyright protections are creatures of American law, and those laws are regularly amended, changed, modified, and updated. After the first copyright law in 1790, Congress made tons of changes in the 19th century to add different types of works to the copyright umbrella, sometimes as technology required (historical prints, dramatic works, photographs, visual art), expanded the protections (including the inclusion of derivative works – extensions of the same original piece), and tweaked the process for securing copyright protection.

Two recent copyright statutes that you may encounter with some regularity in your digital world include:

- **Digital Millenium Copyright Act (DMCA)** (1998) – Briefly, the DMCA was designed to protect creators' work on the internet, making sure people can't easily steal or copy digital content without permission. It gives website owners and artists tools to remove unauthorized content and stop people from breaking copyright rules; and

- **Music Modernization Act (MMA)** (2018) – The MMA was ostensibly designed to improve how music creators get paid in the digital age. In essence, it creates a new system for collecting and distributing royalties from streaming services, making sure songwriters and artists are fairly compensated for their work – whether it does any of these things, of course, is the subject of heated political debate.

An excellent resource for people interested in issues surrounding copyright is, naturally, the U.S. Copyright Office (**www.copyright.gov**). You can read up on the various laws in place (including DCMA and MMA), register your copyright, search copyright records, record transfers of copyrights, and generally overdose on copyright detail. Incidentally, if you enjoy the latter, you might consider law school.

NERD ALERT: All modern copyright law started with the Statute of Anne ("an Act for the Encouragement of Learning"), an English law that became effective in 1710, protecting authors and publishers who registered their new works for 14 years (extendable to another 14 years if the author survived the first 14 – no guarantee in 18th century England, mind you) and 21 years for already published works. Ironically, the Statute of Anne was named after Queen Anne, one of the least known and most forgettable English monarchs of all time. Stated differently, despite it bearing her name, the Statute of Anne sure didn't seem to "encourage" any "learning" about her.

The first copyright act in the United States was passed in 1790. In typical U.S. government fashion, it took several amendments and over a century to get it mostly right. The U.S. Supreme Court didn't hear the first copyright case until 1834 (44 years later!). Indeed, the U.S. Copyright Office was not even established by Congress until 1897 (107 years later!).

What Copyright Protects and What It Doesn't

Not every idea or piece of content qualifies for copyright protection. Just like not every picture, reaction, or notion you have should be shared with the world, even if you are in your underpants (editorial note: sometimes, more is more). Here's a breakdown:

- **Protected**

  o Original creative works fixed in a tangible medium (e.g., a blog post uploaded to your website). The protection is automatic, so you don't need to take a picture of it next to today's newspaper or send it in the mail to yourself, postmarked with today's date – old tricks that people have used. Like most old things, they probably have some value, but they're not worth the hassle, really.

  o Titles, slogans, and short phrases *if* they're creatively styled and may qualify as trademarks (but see below, there is no guarantee).

- **Not Protected**

  o Ideas, facts, or concepts. Compilations of facts are not protectable without some form of creative expression (e.g., telephone book, Grandma's recipe card with just a list of ingredients, words or short phrases like company slogans by themselves).

- o Works not fixed in a tangible form (e.g., an improvised speech).

- o Public domain content (e.g., works published before 1923 in the U.S., or those explicitly dedicated to the public domain).

- o Works not created by a human (or without human intervention) (e.g., selfie taken by a monkey)

- o Pictures of you taken by someone else (that would be a copyright for the person who took it, but as the subject of the photo, you don't get any protection – this becomes an issue with leaked photos, for example – discussed more in Chapter 6).

Example: You've recorded an original song and uploaded it to Spotify. This work is automatically protected by copyright. If someone else downloads it and uses it in their commercial video without your permission, they've infringed on your copyright.

Current Case: In the news right now (around January 2025) is a case initiated by THE NEW YORK TIMES against OpenAI (the maker of ChatGPT) and Microsoft (a major investor in OpenAI) in U.S. District Court (THE NEW YORK TIMES COMPANY V. MICROSOFT CORPORATION, *et al.* – Case No. 23-cv-11195). The case centers on allegations that OpenAI unlawfully used the Times' copyrighted works to train the OpenAI large language models. OpenAI's attorneys filed a motion to compel production of documents related to the Times' use of third-party AI tools, OpenAI's products, and its position on generative AI. OpenAI argued this information was relevant to its "fair use" defense. Mostly the Court rejected these arguments, setting up a fight over what and to what extent the Times can copyright the "news" it prints. Some have called this a huge overreach by the Times, suggesting they are attempting to not only copyright the articles, but also the underlying *facts* in a vain and likely costly attempt to stay relevant in a digital world, especially as their reporters use the same tools they are railing against. Others have applauded the Times' suit as a meaningful check on the advancement of large language models like ChatGPT which utilize work product of others to supplant and replace them entirely. We will we talk a bit more about "fair use" later, but this one is shaping up to be an interesting case.

The Evolving Problem with AI-Generated Content

The question of who owns AI-generated content is quickly becoming the legal equivalent of arguing over who left the lid off the pickle jar. Courts are taking some tentative first stabs at untangling this mess, and, spoiler alert, they're just as confused as the rest of us. Between outdated statutes, overzealous tech enthusiasts, and the undeniable fact that machines won't be issuing takedown notices (yet), the law, as is often the case, is struggling to catch up with the technology.

Take, for example, a recent decision where a court was tasked with determining whether AI-generated artwork could qualify for copyright protection. The answer? A resounding "no." Why? Because the Copyright Act still clings to the quaint notion that creativity requires a "human author."

Yes, apparently, the robot overlord's digital masterpiece is about as copyrightable as the one-liner that killed in the marketing meeting about those losers in accounting. The court's ruling left many scratching their heads and wondering: if machines can do half (or all of) the work, do humans even deserve credit?

The courts aren't entirely heartless (or clueless). Judges are starting to dig into the nuances of human involvement. If you're actively (and there will be cases about what this means in the future) directing an AI's output—choosing prompts, tweaking settings, and curating results—you might have a fighting chance at claiming copyright.

But let's not kid ourselves; telling Adobe Firefly to draft a cool butterfly picture is not "active" in any real sense of the word. "But, your honor, I told it to make it purple and cool." That approach is not likely to cut it.

Ownership over purely AI-generated content is being treated like Bigfoot sightings: a fun concept, but (maybe) ultimately a myth. Frankly, even Adobe isn't sure what do about commercialization of their AI-generated stuff. Right now, they're basically say, "go ahed and use it." That may change some day.

For creators navigating this legal thunderdome, the safest bet is to stay involved. Courts want to see that good, old-fashioned human touch. If you want to slap your name on that AI-assisted masterpiece, make sure you're doing more than just hitting "Generate." While the law grapples with the implications of artificial creativity, one thing is clear: the fight over AI copyrights is just getting started, and it's already a glorious dumpster fire.

How Copyright Attaches Automatically (But Registration is Better)

Copyright protection is automatic as soon as a work is created and fixed in a tangible form. There is no magic symbol that needs to be attached to the work. The © symbol gives the world notice that you have copyrighted the work, but it's not necessary to have copyright protection apply. That said, use it whenever you can to avoid the argument by an infringer that he or she "didn't know" your work was subject to copyright. I'll discuss the registration process below, but it's basically a formal filing with USPTO of your copyright. You don't need to register your work to have copyright protection, but registration comes with significant benefits:

- **Legal Protections:** Registered works can be enforced in court, allowing creators to sue for damages and attorney's fees in cases of infringement (if you don't register, you can't sue to stop another's use of your copyrighted works – don't worry, you can register even after you find out – see below for more information);

- **Public Record:** Registration serves as a public record of ownership, deterring potential infringers;

- **Proof of Ownership:** A certificate of registration is a strong form of evidence in disputes over ownership or originality; and

- **Eligibility for Statutory Damages:** Without registration, only "actual" damages can be claimed (and those must be proved with evidence – often require expert testimony and can be quite difficult to prove). Registration opens the door to statutory damages, which can be substantial (and fall under a different standard of proof – i.e., the damages are presumed and can be like $150,000 per violation). Statutory damages can also include the legal fees you had to spend to enforce your rights, so don't sleep on that benefit. Lawyers are expensive. We can charge $150,000 in a month (no lie).

How to Register Your Copyright

The process of copyright registration in the U.S. is straightforward:

1. **Determine Eligibility.** Ensure your work qualifies for copyright protection. It must be an original creation and fall under one of the protectable categories such as literature, music, art, or software.

2. **Complete the Application.** Submit an application through the U.S. Copyright Office website. You'll need to provide details about the work and its creator(s).

3. **Pay the Fee.** Registration fees vary depending on the type of work and filing method, but they are typically modest (meaning, not too expensive). You're looking at something like $50.00 to avoid huge costs later.

4. **Submit a Copy.** Provide a copy or sample of the work being registered. This copy will become part of the Copyright Office's records.

When and Why to Register Your Work

As I said before, it's best to register your copyright as soon as your work is completed and fixed in a tangible form. Under U.S. copyright law, copyright protection begins automatically as soon as an original work is created and fixed in a tangible medium (such as written text, recorded music, or a digital file). However, there are important distinctions between unregistered and registered copyrights:

Protections Without Registration

- **Automatic Rights.** You retain exclusive rights to reproduce, distribute, perform, display, and create derivative works from your creation.

- **Limited Legal Recourse.** While you can stop others from infringing on your work, you may not file a lawsuit to enforce these rights unless the copyright is registered.

Late Registration is Possible

Even if you choose to register your work after infringement has occurred, you can still secure some protections:

- **Actual Damages and Profits.** You can seek actual damages, which are based on the financial harm caused by the infringement, and any profits the infringer made from using your work.

- **Prospective Protection.** Registration provides additional tools to deter future infringements and strengthens your ability to enforce your rights going forward.

Drawbacks of Delayed Registration

- **Loss of Statutory Damages.** If your work is not registered before the infringement occurs (or within three months of creation), you cannot claim statutory damages or attorneys' fees. These damages can be significant and are easier to prove than actual damages.

- **Weaker Evidence.** Early registration serves as *prima facie* evidence of ownership, making it easier to prove your case in court. Delayed registration may result in the need for more extensive evidence to establish your rights.

Let's sum up the general rule: Once you create something in tangible form, it has copyright protection. You should register it with the U.S. Copyright Office (but, if you don't, it's not the end of the world). Keep an eye out for infringing uses by others. If you see some bastard using your designs, call a lawyer right away. If you haven't registered your work, go register it ASAP. You might still be on the hook for your attorneys' fees, but you can shut the infringing bastard down and sue him for actual damages. Short story, even shorter:

- **Registered?** Statutory damages, actual damages, and/or attorneys' fees, plus all of the benefits of unregistered protection.

- **Unregistered?** Actual damages available for the infringement IF you register the copyright after infringement. No attorneys' fees, no statutory damages, but you can sue once you register to stop the other side from ripping you off.

So, this has all given you a sense of what you can protect, how to do it, and how to stop infringement if you see someone else using your material. Now, let's look at the single biggest defense your opponent will plan on making: "Fair Use."

Fair Use: A Critical Exception

Fair use allows limited use of copyrighted material *without* permission, under specific circumstances. Think of the Daily Show. It's satire and parody, looking to skewer the news with funny takes on statements made by politicians, celebrities, and others. Pulling down a snippet of a Donald Trump press conference and making commentary or imitating the tone/language, well, that built an empire and a career for Jon Stewart. It's also the best example of the "fair use" defense. When used, "fair use" is commonly invoked in cases involving:

- Commentary or criticism (e.g., reaction videos).

- Education and research (e.g., classroom use of materials).

- News reporting (e.g., including excerpts of speeches and other copyrighted material to report on events).

- Parody and satire (e.g., comedic reinterpretations of music videos).

- Writing quasi-funny books about copyrights and mentioning certain examples.

Look, human nature being what it is, everyone who asserts a "fair use" defense is going to try to shoehorn their excuse into some category of protected use. They will have lawyers; you will have lawyers. Hindsight is 20/20 when it comes to litigants' "intentions" with unauthorized use of content, meaning that those "intentions" will be whatever their lawyer tells them they were.

LAWYER: Were you attempting to satirize

the content?

INFRINGER: Oh, yeah. Let's go with that. What does satirize mean? I mean, of course, never mind; I was totally into satire.

Let's take it out of the parties' hands and let's see what courts consider when it comes to determining whether something qualifies as "fair use":

1. **Purpose and Character of Use.** Is it transformative (i.e., adds new meaning or value)? Nonprofit, educational, or personal use is more likely to qualify as fair use compared to for-profit uses.

   - *Example:* A video essay analyzing a film's themes with short clips might qualify as fair use.

2. **Nature of the Work.** Published works (that you use) are more likely to fall under fair use than unpublished ones. Likewise, factual works, such as news or academic articles, are more likely to fall under fair use than highly creative works like novels, music, or films.

   - *Example:* Using a public speech for educational purposes is more likely fair use than leaking an unpublished manuscript.

3. **Amount and Substantiality.** The smaller the excerpt used, the more likely it's fair use.

- *Example:* A meme using a single frame of a movie might be fair use; an entire video clip likely isn't. BUT: Even a small portion may not be fair use if it constitutes the "heart" of the work (e.g., the most memorable scene of a movie or a critical plot point in a book). Spoiler alert: this means leaking spoilers as part of your use are often not considered fair use. Umm, The Crying Game twist at the end? That's one well-tread example. Also, The Usual Suspects. Coincidentally, Gabriel Byrne was in both of those movies. Coincidence? I don't think so.

4. **Effect on the Market.** Does the use harm or damage the market for the original?

- *Example:* A creator reposting full, unaltered episodes of a TV series undermines the original's market and wouldn't qualify as "fair" use. The effect is that your unauthorized use is essentially substituting itself in the market for an authorized use.

Fair use litigation is brutal and can take years to sort out. The primary defense by OpenAI and Microsoft in the case mentioned above is going to be that any use by ChatGPT (if there was any) would qualify as fair use and, as such, the Times is just making much ado about nothing. There are interesting arguments to be made on both sides. This will last for years because, even if your use arguably meets all of the factors that might suggest a fair use, it's not automatic. Courts will look at the factors holistically and, ultimately, make a judgment call. May be right, may be wrong. But, the court will eventually decide.

As a litigator with 20 years of experience, I question the value of the Times fighting this fight for five or so years, especially given the minimal impact a "win" will have on its bottom line. If they win, what does that mean? Doesn't generate new readers, damages may be impossible to prove, and they look like a dinosaur roaring at the meteor falling from the sky, no matter what. If I'm the lawyer for the Times, however, I know it means I'm buying a new boat, win or lose. So, that's nice.

NERD ALERT: What about "Weird" Al Yankovic, you may be asking? He's made parody videos of huge artists (Gerardo notwithstanding) that have been shown the world over. You might have seen "Eat It" (a parody of Michael Jackson's "Beat It"), "Taco Grande" (a parody of Gerardo's "Rico Suave"), or "Smells Like Nirvana (a parody of Nirvana's "Smells Like Teen Spirit"). He copies the music, often the music video, and changes the lyrics around to make it funny (arguable). How can he get away with that? Does he need permission from the artists?

The answer is: **No. He doesn't need permission.** *This is the biggest exception to copyright – parody/satire.*

Weird Al claims to have a personal rule, however. He has chosen to not make parody songs by artists who don't approve. Prince was, apparently, a notable artist who could not countenance Weird Al making fun of his terrible music. It's okay. The rest of society does it for him. Al simply moved on. Did he have to? Absolutely not. He has made a business decision that he chooses to self-enforce. From a strictly legal standpoint, however, satire and parody are fair use and there's not anything Taylor Swift or Sabrina Carpenter can do about it.

Personally, I came up with a much better version of Dua Lipa's "Levitating" a few years back. I called it "Rehydrating." I didn't release it because of my own personal rule of not distributing videos that will sink my business, embarrass my family and myself, and make no money. It's a pretty good rule.

Examples of Copyright Issues in the Creator World

1. **Music in Videos**

- A Twitch streamer used copyrighted music in their background playlist. The content was flagged, and their monetization was suspended. To avoid this, creators should use royalty-free music or licensing platforms like Epidemic Sound.

2. **Image Misuse**

- A popular YouTuber used a stock image without checking the licensing terms. Some terms allow you to use an image for personal use, but require a different license for commercial use – or prohibit commercial use. The image's owner in this example demanded $10,000 in damages. Licensing agreements, like all contracts, will govern the rules the parties play by. We'll discuss contracts and agreements in more detail later, but don't assume that "stock" assets can be used for any and all purposes.

3. **Reaction Videos**

- A creator made reaction videos featuring extended clips from copyrighted movies. While transformative content is more likely to qualify as fair use, extensive use of the original work led to takedowns. The creator was forced to restructure their content, reduce clip length, and add more original commentary in order to avoid further takedown demands.

Practical Challenges Ahead: The real difficulty is policing the staggering amount of content on the internet. What kind of use of a background song is actionable with someone dancing over the track? How long is too long? Is it covered by platform licenses with music labels? Is it not? Do you want to spend a lot of legal dollars to figure it out – and maybe lose?

There are no solid answers to these questions and exceptions for every rule. The chaos creates more "regulations" but less "compliance" because compliance is nearly impossible. A more pragmatic approach seems to be taking over the content creation space where creators are cognizant of the spirit of the rules and are giving attribution to songs, but that's about it. There's no licensing happening. They are creating so much varied content that individual takedown requests effectively mean nothing (i.e., no particular submission is monetizing the whole account), the exposure of your music (even if unauthorized) might be more valuable than any legal fight, and sophisticated creators have so many backup accounts that demonetizing one account only drives people to other ones (a "whack-a-mole" problem).

In a world where every trip to Starbucks has a backing track and every outfit change has another, copyright owners (especially music creators) might be best served to ride the wave and take all the exposure they can get, focusing on deals with platforms to contribute to licensing collectives. This could ultimately erode the "fair use" exception, but that has not been tested thoroughly in court just yet. Music seems particularly amenable to platform-level licensing. Where this will get weird and difficult is when AI-generated versions of people's likenesses start creating entirely unaffiliated accounts, some of them funny, others defamatory or pornographic. When that happens, the outrage will be justifiable, but hopefully enforcement (or lack thereof) of the existing rules won't have been so watered down that even protecting your own likeness becomes impossible.

If there's one thing that is as certain as death and taxes, it's that content creators will not stop making content, with or without your permission. For every high-profile account, there are literally thousands that are doing the exact same thing under your radar. You might tag a few of them here or there, but, until they have real assets to lose, they're not slowing down, let alone stopping.

Copyright Claim Systems and How to Navigate Them

When you're looking to enforce your

copyrighted content, there are some tools available. Platforms like YouTube and Instagram have automated copyright claim systems to detect and address potential infringements. Here's how to deal with them:

1. **Content ID on YouTube**

   - *Pros*: Automatically detects copyrighted material in uploads.

   - *Cons*: Often flags legitimate fair use content.

   - *Tip:* If you believe your content is transformative and qualifies as fair use, file a dispute. Provide clear explanations and timestamps to support your claim. Don't expect this will just sort itself out in 10 minutes, either. You may want to have a lawyer do the arguing and file the dispute.

2. **DMCA Takedowns**

   - *Reason*: Copyright owners can issue takedown requests under the Digital Millennium Copyright Act (DMCA) if they feel someone is infringing their copyrighted material or using it in violation of a valid license.

   - *Process:* Short but sweet (to explain), this is the usual process:

1. **Notice to Platform** - Copyright owner issues takedown notice to the platform (e.g., Facebook, Instagram, etc.). This notice must include specific elements: the copyright owner's signature, identification of the copyrighted work and infringing material, contact information, and statements of good faith belief in infringement and accuracy under penalty of perjury;

2. **Removal & Notice to Infringing Party** - Upon receiving a valid notice, the service provider typically removes the content promptly and notifies the alleged infringer;

3. **Counter-Notice** - The accused party then has the option to file a counter-notice if they believe the takedown was unwarranted. If no counter-notice filed, the platform will likely assume the use was an infringement and wrongful (see consequences below);

4. **Restoration of Content and/or Potential Suit** - If a counter-notice is received, the service provider must restore the content within 10-14 days unless the copyright owner files a lawsuit. This will

hopefully shake out the "strategic" filers, meaning the ones who are jealous and are just trying to mess with you. They won't file a lawsuit.

3. **Watermarking and Metadata**

- Embed watermarks and metadata in your work to assert ownership and deter infringement. Tools like Photoshop and Lightroom include metadata options. If you don't think your racy Instagram post is being monetized, take a walk down Las Vegas Boulevard sometime and collect the free cards being handed out. You might see a familiar face...

NERD ALERT: Let's talk about the implications of DMCA takedown notices. What are the consequences for creators? Platforms typically implement a "strike" system when dealing with multiple takedown notices for content from the same user. Here's how they generally handle such situations:

- *Warning System Account holders usually receive a warning or "strike," for each infraction. This serves as a formal notification of the copyright violation.*

- *Account Termination After a series of infractions, typically three strikes, service providers will terminate the user's account. This "three-strike" policy is common across many platforms.*

- *Account Demonetization   The platform can remove the ability of a user to monetize its content, withhold funds that have been earned, or, as above, terminate the account entirely.*

- *Content Removal   Upon receiving a valid DMCA takedown notice, platforms typically remove the infringing content promptly. This is done to maintain their safe harbor protections under the DMCA.*

- *User Notification   The platform usually notifies the user responsible for the infringing activity about the takedown and the potential consequences of repeated violations.*

- *Repeat Infringer Policy   Platforms are required to implement and enforce a repeat infringer policy to maintain their DMCA safe harbor protections.*

- *Legal Consequences   In severe cases where users knowingly and willingly post copyrighted material multiple times, they may face criminal penalties and lawsuits.*

Platform rules change, people use DMCA notifications to undermine competitors, and things can get missed, so it is crucial to monitor your compliance with platform rules and do not ignore notices. If a particular platform is the lifeblood of your yearly earnings, well, you may be well-served by spending some money on a lawyer to protect your interests. It's never easy to tell when you need legal help in this kind of situation, but, generally speaking, the earlier the better.

Lawyers don't have time machines and "too late" is, unfortunately, "too late."

Licensing: Sharing Your Work on Your Terms

Licensing allows creators to permit others to use their copyrighted material under specific conditions. Types of licenses include:

- **Exclusive License.** Grants rights to a single licensee.

- **Non-Exclusive License.** Allows multiple parties to use the work.

- **Creative Commons.** A set of free, standardized licenses allowing creators to specify how their work can be used (e.g., for non-commercial purposes only).

Example: A creator licenses their music to a podcast producer for a one-time fee. The agreement specifies that the music can only be used for that podcast episode or episodes, protecting the creator's broader rights.

We will get into licensing in more detail later in the book. For now, just know that there is value in protecting your copyright, even if it doesn't seem like it early on in your career.

Practical Steps to Protect Your Copyright

1. **Register Your Work.** Especially for high-value creations or those prone to misuse.

2. **Understand Licensing Terms.** Before using third-party content, review the associated licenses.

3. **Monitor Infringement.** Tools like reverse image search and YouTube's Content ID can help track unauthorized use.

4. **Create Clear Policies.** If you license your work, use contracts that define usage terms, duration, and compensation.

5. **Educate Yourself.** Familiarize yourself with copyright laws and platform-specific guidelines to stay compliant.

By understanding the essentials of copyright, you'll not only protect your creative assets but also confidently navigate the complex landscape of online content creation when, at the drop of a hat, you can be accused of infringing on the rights of others. Copyrights are both a sword and a shield.

Next, we'll explore the ins and outs of trademarking, ensuring your brand stands out in the crowded digital space.

CHAPTER 3

Trademark Essentials

What Are Trademarks?

Trademarks protect the identifiers of your brand, such as names, logos, slogans, and even unique colors or sounds, ensuring they distinguish your work from that of others. For content creators, trademarks can apply to:

- Your channel or brand name (e.g., "Cooking with Casey").

- Your logo or visual branding elements (e.g., the Nike swoosh).

- Catchphrases or taglines unique to your content (e.g., ring announcer's "Let's Get Ready to Rumble!").

Unlike copyrights, trademarks *require* registration to gain full legal protection, though some rights are acquired through mere use. While there may be some common law trademark rights for unregistered marks used in commerce, there is no "automatic" protection for trademarks. If you are on the fence and you want to really turn this into a monetizable brand, you should attempt to register to be safe.

Why Trademarks Matter for Content Creators

As a content creator, trademarks are critical for:

1. **Building a Recognizable Brand.** A unique name or logo helps your audience identify and remember you.

2. **Protecting Against Copycats.** A registered trademark gives you legal recourse against others trying to imitate your brand.

3. **Expanding Monetization Opportunities.** A trademarked brand can be used on merchandise, sponsorships, and licensing deals.

Simple Example: A gamer named "SuperDingus" trademarks his name and logo. When another creator starts selling t-shirts using the same name, SuperDingus can enforce his trademark to stop the infringement. He can also demand a licensing deal out of the infringing party, likely with some share of the past sales money on merch using the trademark.

Trademark Basics: What Can Be Trademarked?

The scope of trademarks goes beyond names and logos. Here's a closer look:

- **Names.** Your business or channel name, provided it is distinctive (e.g., "VlogVibes" is more trademarkable than "Fashion Vlogs"). When in doubt, deliberate misspellings are more likely to suggest a distinctive use than ordinary language. A company's mark of "BBQ Jerkz" is definitely more distinctive than "Barbecue Jerks."

- **Logos.** Any unique graphic design representing your brand. An excellent example of this is Coca-Cola, which has registered its distinctive lettering, bottle shape, and standard character formats. Think it only applies to one color? Nope.

- **Slogan.** Catchy phrases associated with your content (e.g., "Git-R-Done," "Make America Great Again").

- **Trade Dress.** The unique visual look and feel of your brand, such as packaging or website design (e.g., the Louboutin red sole, the red sealing wax on Maker's Mark).

- **Sounds.** Audio trademarks, like Netflix's iconic "ta-dum" sound, the MGM lion roar, and the NBC chimes.

Practical Tip: Generic or non-descriptive terms (e.g., "My Travel Log," "Apple Mail") are harder to trademark unless they develop exceptionally strong brand recognition.

The Curious Case of Stephen He

13.3 million YouTube subscribers. 7.7 million TikTok followers. 3.2 million Instagram followers. Numerous viral catchphrases. ZERO registered trademarks?!

Anyone with a 10 year-old in earshot has been regaled with a non-stop chorus of "emotional damage!" and "what the hail you say?" Even the "Beijing Corn" ads and bits have become ubiquitous and (weirdly) quotable. So, why has the US-based Chinese-Irish comedian not protected any (or all) of this viral gold with trademark registrations? It's not entirely clear, but there could be a number of reasons:

- The trademark process takes time and is not guaranteed. In a fast-moving digital world, getting the registration months later may be too little, too late. No one may care about "Beijing Corn" a few years from now. By that time, Mr. He may have simply moved on to new ventures or catchphrases.

- It could just be a business decision. Run-and-gun and have fun, perhaps. Maybe he's making very little from merchandising when compared to his platform advertising revenue. And, maybe the viral nature of his marks are reward enough. As a lawyer who has practiced for two decades, I stopped judging clients years ago about their decisions – my job is to handle consequences.

- He may have investigated it and found the case too close to call, making it an expensive "failure" if it doesn't work out for him. People could make arguments both for and against his catchphrases being trademarkable at all. They are, at the same time, distinctive and not. At the same time, iconic and ordinary. There's no specific font, logo, or distinctive imagery that swings them into the "definitely registerable" camp. Discretion is the better part of valor, sometimes.

Personally, fickle though catchphrases and influencer popularity may be, were I his attorney, I would certainly advocate for the implementation of a trademark registration strategy if only to preserve opportunities that may not have materialized just yet. You just never know how this stuff makes its way back around (the 90's are back, y'all). I'd rather be safe than sorry, especially given his massive following. Plus, I wouldn't mind a free "emotional damage" shirt.

The good news is that he (He) probably has some protection with common law trademark rights where he does business (e.g., sells coffee mugs, drink powders, t-shirts, etc.), even for marks that may not be registerable with USPTO. That may be enough to keep the shameless copycats at bay, at least.

No matter what, Stephen He has transformed the landscape of content in ways I'm not sure anyone saw coming and, if nothing else, thanks for the laughs, dude. If you need a lawyer, hit me up.

How to Register a Trademark

Trademark registration varies by country, but the process generally includes:

1. **Conducting a Trademark Search**

   - Use tools like the USPTO's Trademark Electronic Search System (TESS) to check for existing trademarks. In case you want to play around, here's the link: **https://tmsearch.uspto.gov/search/search-information**. Try some simple searches. Use different combinations, spellings, punctuation, etc.

- *Example:* A creator named "Gadget Guru" discovers that their name is already trademarked for tech reviews. You can save a lot of pain up front with a simple (free) search.

2. **Filing an Application**

- Submit an application to your country's trademark office (e.g., USPTO in the U.S.).

- Include details about the mark, its usage, and the categories of goods or services to which it applies (e.g., clothing, food, industrial goods, etc.).

3. **Meeting Usage Requirements**

- In most cases, trademarks must be *used in commerce* before final registration. As above, if Stephen He is just making new videos and getting new laughs but not "commercializing" those catchphrases, that's not good enough to satisfy registration requirements.

- *Example:* A fitness creator uses their brand name "SweatPro" on workout gear before securing the trademark.

4. **Maintaining and Renewing Your Trademark**

- Trademarks must be renewed periodically (e.g., every 10 years in the U.S.).

- Failure to renew could result in losing your protection.

Common Trademark Challenges

Creators often encounter the following issues when dealing with trademarks:

1. **Name Conflicts**

- Choosing a name already in use or already trademarked can lead to legal disputes.

- *Example:* A vlogger named "Epic Adventures" receives a cease-and-desist letter because a travel agency already holds a trademark on that name. A simple TESS search could have avoided a mountain of headaches.

2. **Descriptive Marks**

    - Using overly generic names makes it difficult to obtain trademark protection.

    - *Example:* A podcast titled "True Crime Stories" struggles to secure a trademark due to its generic nature.

3. **Enforcement Costs**

    - Protecting your trademark can require significant time and money. Move quickly and hit hard. Bullies don't respond to legal letters or to pleas to the better angels of their nature. You need to punch them in the mouth. Unapologetically.

    - *Example:* A creator spends months pursuing a legal case against a merchandiser using their logo without permission. There's no guarantee that merchandiser doesn't just declare bankruptcy or disappear 6 months into the case, leaving you with no recourse and hundreds of thousands of dollars in legal bills. Real litigators with real skills know how to hit them in different ways (e.g., shut down their bank accounts, merchant agreements, etc.) and squeeze them of the benefits of their ill-gotten gains before they get a chance to screw you a second time. Find one of those lawyers. Maybe one who has written this very book...

Real-World Trademark Disputes

1. **The Fine Bros vs. the Internet**

    - The Fine Bros attempted to trademark the term "react" with regard to their reaction videos (as well as "Adults React," "Teens React," and "Kids React"), introducing something they titled "React World." Ultimately, their plan was to co-opt an entire YouTube genre by demanding licensing of similar videos to create a monopoly of sorts. Some might call that approach a "shake down." Others might call it "just good business." But, those people who suggested the latter were, well, idiots.

    - The brothers lost hundreds of thousands of followers in a matter of days, with some estimates of 600,000 or more in total.

    - The move sparked backlash, as many argued it was too generic and stifled creativity.

Maybe, more importantly, users recognized the shameless ploy to profit from a viral trend by opportunists who had too high an opinion of their relative worth on the internet.

- The Fine Bros eventually withdrew their applications, issued an apology, and ceased all enforcement efforts against their fellow content creators. It didn't stop a nosedive in popularity or dozens of former employees emerging with claims of sexist and racist discrimination claims against the brothers in the years that followed, with basically no one having heard from the Fine Bros since 2020.

- From internet darlings to anonymous nobodies in the span of less than 5 years. Rough.

2. Kylie Jenner vs. Kylie Minogue

- Kylie Jenner's attempt to trademark "Kylie" was challenged by singer Kylie Minogue, who argued (successfully) it would cause confusion.

- The case highlights the importance of ensuring your trademark doesn't infringe on established marks. A simple TESS search (again) could have avoided this nonsense.

- It's the ultimate type of hubris to assume what you are doing now (Jenner) is more important that what has come before (Minogue). As expected, there were some catty fireworks, including Minogue's lawyers describing their client as an "internationally-renowned performing artist" while describing Jenner as a "secondary reality television personality." Meow and ouch. Not for nothing, but can anyone name the last popular song Kylie Minogue recorded? Any song Kylie Minogue recorded?

- Jenner's application was rejected by the USPTO. That said, she would eventually console herself with a re-branding of her cosmetics line and about a billion dollars generated from that new brand. Her status as a "secondary reality television personality" is still a valid point.

3. **Instagram Influencer Names**

- A fitness influencer discovered that a competitor trademarked a similar name, forcing them to rebrand. Early trademark registration could have prevented this.

- Simple achievable, disciplined steps to make for an overall healthier product (or trademark)? One would think a "fitness influencer" would have understood the concept. Um, that's kind of the point of their existence.

- That said, Only Fans is littered with "fitness influencers" who know nothing about fitness, trademarks, or discipline.

Protecting Your Trademark

Let's say you have gone to the trouble to get a trademark. That's not the finish line. Trademarks are only as strong and as valuable as the effort you put into protecting them. Here's how:

1. **Monitor for Infringement**

   - Regularly search for unauthorized use of your brand's name, logo, or other marks. Tools like Google Alerts can help, as can newer AI-based search engines (e.g., Perplexity).

2. **Enforce Your Rights**

- Seek out legal help. Consultations are usually free with most lawyers. Once you know the potential costs and benefits of enforcement efforts, you can make the decision to proceed or not.

- Send cease-and-desist letters or pursue legal action when necessary.

- *Example:* A creator finds their logo being used on counterfeit merchandise. They issue a takedown notice to the seller, to the platform, and look to disrupt their activities in other ways (see above re: seeking out legal help).

3. **Educate Your Audience**

- Make your trademarks visible and clear. Add ® or ™ to your marks where appropriate.

  o The Registered ® Symbol – this mark next to your brand means the trademark has been registered by USPTO. This is the best mark you can get. But, if you just tack it on without having received a registered trademark from the USPTO, it's a violation of Federal law and, if you have an application pending while using the mark inappropriately, expect it to be rejected. This mark serves as notice to the whole world that your trademark is registered; infringers should beware.

o The ™ Symbol – this mark next to your brand doesn't really mean much of anything except that you consider the brand to be something you intend to protect. Doesn't mean there is a trademark or even an application for one. But, frankly, it's better than nothing, so use it. In the U.S., you'll also see an SM symbol (which stands for Service Mark). This is usually applied to marks used in connection with the provision of services (e.g., accounting services, personal training, legal services, etc.) as opposed to a "good." Most other countries don't bother with the distinction. Much like the ™ symbol, it doesn't indicate much of anything legally. Practically, it suggests you know what a service mark is and that you are probably going to enforce it, if need be. So, again, just use it.

Licensing and Merchandising with Trademarks

Why all the bother, really? Short answer, cash

money. Trademarks can create new revenue streams through licensing and merchandising:

1. **Licensing Agreements**

   - Allow others to use your trademark for a fee or royalty. We will discuss this more in later chapters around licensing, but the rule of thumb is ridiculously simple: a) define the terms clearly, b) in a contract, and c) before any potential use.

   - *Example:* A lifestyle creator licenses their logo to a clothing brand for an exclusive line of hoodies. Sometimes, these agreements are styled as "collaborations" or, in the parlance, "collabs." Essentially, it's just a licensing agreement to use your brand or mark on their stuff. These can be long-term, short-term, one-off, or continuing, but they usually are quite lucrative if your following really wants a new hoodie.

2. **Merchandising Opportunities**

   - Use your trademarks on branded products like t-shirts, mugs, or planners.

   - *Example:* A podcast called "Mindful Minute" might sell trademarked journals and meditation guides. "Uncle Roger" (Nigel Ng) has all sorts of branded goodies and gadgets (my favorite is the "Haiyah!" button). Podcasts can and often do generate a ton of revenue from merch sales. Some people even write books to capitalize on their trademarks and branding...

Practical Steps to Secure Your Brand

1. **Choose a Strong, Unique Name.** Avoid generic or descriptive terms and aim for something memorable (or at least distinctive).

2. **Register Early.** The sooner you file for a trademark, the easier it is to prevent others from using your brand.

3. **Use Your Mark Consistently.** Ensure your branding is uniform across platforms to strengthen recognition. Every time you change your brand, you have to start again.

4. **Seek Legal Help.** For complex filings or disputes, consult an attorney. Good practitioners have a network of attorneys to refer you to. These disputes can be very intricate and may require the expertise of someone who specializes in that area of law. Start with any lawyer, finish

with the one that kicks ass regularly in the area you need.

By understanding the power of trademarks and taking proactive steps to protect them, you'll solidify your brand's place in the creator economy. It may also net you some sweet, sweet cash if you do it right (I'm talking to you, Stephen He). If you think you're smarter than people who specialize in this stuff and try to do it yourself or mess it up, okay. Dumb decisions are made every day. It happens.

But, fair warning, you might just end up like the Fine Bros, watching your multi-million dollar empire blow up in your face, big-time. You'll be asking people if they "want fries with that" in the drive-thru sooner rather than later.

Just like copyrights, trademarks can be a valuable asset (intangible though they might be) that gives you leverage in negotiations, a foundation on which to monetize your good ideas, and a hammer to use if someone tries to rip you off.

Next, we'll explore patents and trade secrets, diving into their relevance and potential applications for content creators.

CHAPTER 4

Patents & Trade Secrets

Patents and trade secrets might seem like they belong more to the realm of scientists and inventors than to YouTubers, Instagram influencers, or digital brand ambassadors. But, these two areas of intellectual property can have surprising relevance for content creators. Let's break it down and see how they can protect your creative and innovative efforts—and maybe even give yourself a competitive edge.

Patents: Protecting Your Big Ideas

A patent protects inventions or processes that are novel, useful, and non-obvious. For most content creators, patents might not immediately come to mind, but they could apply in several unexpected ways.

DISCLAIMER: Patent lawyers and specialized IP lawyers have a technical or scientific background, a USPTO authorization, and have passed the Patent Bar examination. I have (mercifully) done none of those things. While I have practiced commercial law for 20 years, including overseeing patent prosecutions and litigation by patent lawyers, I don't profess to be an expert in this specialty area. This makes me just smart enough to be dangerous, like a teenager with access to the Door Dash account and minimal parental oversight. If you need an expert, let me know. I will find you one. Depending on the state of the economy, maybe even one who delivers Taco Bell for Door Dash as a side-hustle.

What Can Be Patented?

Patents can cover:

1. **Physical Inventions.** If you've invented a piece of equipment to improve your content creation process, like a unique camera rig or a lighting setup, a patent might be worth considering.

2. **Software Innovations.** Developed an app or plugin that automates some part of your workflow? Software patents could come into play.

3. **Creative Processes.** Even unique methods for editing or delivering content could potentially qualify.

Example: Let's play the "theoretical" game. Imagine you're a gaming creator who develops a tool that allows fans to interact with your livestream in real time through custom emojis and animations. This type of innovation could be patentable, giving you exclusive rights to monetize it. All of this is "theoretically" possible, but as you learn more below about the process of getting a patent, you might want to dial back that enthusiasm, software nerd.

Types of Patents

1. **Utility Patents.** These protect how something works. If your invention does something new and useful—like a gadget or software—this is the patent you'll want. Think flying car or monkey butler.

2. **Design Patents.** These protect the way something looks. If you've created a uniquely designed product (like a custom microphone with a futuristic aesthetic), you might consider a design patent.

3. **Plant Patents.** Not relevant unless you're a gardening-focused creator who has bred a new type of tulip... in which case, congrats. I guess.

Steps to Patent Something

1. **Determine Patentability.**

   - Your idea must be new and not "obvious." This means doing a thorough patent search to make sure no one else has already claimed it. As with trademarks, start at the USPTO (**https://www.uspto.gov/patents/search/pat ent-public-search**).

   - *Tip:* You can also check out free tools like Google Patents to assist with your search efforts. Best bet, though, is to consult a patent attorney for more robust searches.

2. **Prepare Your Application:**

   - Be detailed! Your application needs to include drawings, descriptions, and claims about what your invention does. You <u>DO</u> need to disclose the "secret sauce" and tell the world how to do exactly what you have figured out. Sounds counterintuitive, right? That's true, but it's also the price of getting a patent. You could just monetize your idea and hope no one figures out the "sauce" – that's also an option that a good patent lawyer should discuss with you.

   - *Example:* A creator patents a modular green screen that can be adjusted for different shooting angles. Their drawings and descriptions need to show how it works in detail.

3. File With the Patent Office

- In the U.S., this means filing with the USPTO. Other countries have similar agencies.

- Be prepared for a lengthy review process, weird revision requests, and lots and lots of fees.

4. Enforce Your Patent

- Once you have a patent, it's yours to use exclusively – even though everyone else already knows how to replicate it. Utility patents are good for 20 years, design patents for 15, and plant patents for 20, as well. These terms are measured from the filing date of the patent.

- If someone copies your invention, you'll need to step up and protect your rights— which can mean anything from a friendly letter to full-on legal action.

- *Note*: failing to protect your patent (i.e., sitting on your rights) could mean you lose those rights, so be ready to move if you see infringement. No point spending all of that money on securing a patent only to have it be worthless because you sat on your hands.

Downsides to Patents

As above, patents can be ridiculously expensive and time-consuming to obtain. There are various estimates out there, but the general consenses appears to be roughly 24 months from the time you file an application until you (hopefully) receive a patent. They're best for creators who genuinely have something unique that's worth protecting—and who have the resources to defend it, if necessary.

Unlike the other forms of IP we discussed (i.e., trademark and copyright), the decision to go down the patent rabbit-hole is not for the faint of heart. And, patent lawyers don't take cases on contingency, either. You will pay through the nose. It will take a long, long time. And, you will likely be disappointed at the end. Even worse? You've just told everyone on the planet how to build your invention. Yikes.

When it comes to enforcing patents, the ramifications can also be extremely severe. It is not unheard of to see this scenario play out: 1) Patent owner gets patent from lazy examiner in the USPTO; 2) owner sues another party for infringement; 3) the alleged infringing party defends that the patent should never have been issued in the first place; 4) the court agrees; and 5) the patent is invalidated!

Trade Secrets: Silent, but Deadly

Unlike patents, trade secrets are not registered. Instead, they rely on (and require) keeping certain information confidential. A trade secret could be anything that gives you a competitive advantage and isn't publicly known. If you've figured out how to game the UBER algorithm to get cars to you faster at a lower trip price, keep it under your hat. You can charge people to make it work for them, obviously. If you tell them how it works or how to teach their friends and family to do the same thing, you just lost your trade secret, genius.

What Qualifies as a Trade Secret?

To qualify, a trade secret must:

1. Be commercially valuable.

2. Not be generally known.

3. Be protected by reasonable measures to maintain secrecy.

For content creators, trade secrets might include:

- Your video editing techniques.

- A unique process for scripting or brainstorming.

- Monetization strategies, like a secret formula for going viral.

- Proprietary workflows, like a detailed process for collaborating with sponsors.

Example: A cooking creator has a secret method for making viral recipe videos, involving specific angles, lighting setups, and timing for food reveals. As a result, they can generate posts 5 times as fast and possess the ability to generate multiple, engaging posts from the same footage. By keeping this process under wraps, they maintain a competitive edge in the crowded food content space.

Protecting Your Trade Secrets

1. **Limit Access.** Only share trade secrets with people who need to know, and even then, use non-disclosure agreements (NDAs) (also called confidentiality agreements).

    - *Example:* A team of editors working for a creator signs NDAs to ensure they don't reveal the creator's unique editing tricks.

2. **Document Everything.** Keep records of your trade secrets, including how they're used and the measures you take to protect them.

3. **Be Vigilant About Security.** Use password protection, encrypted files, and other digital security measures to safeguard sensitive information.

4. **Enforce Your Agreements.** There's no point going to all of the trouble to get people to sign NDAs if you're just going to ignore them when they are violated. Almost all NDAs have injunctive relief provisions which allow you to run immediately to court to stop (i.e., enjoin) the other party from sharing your trade secrets or using them to compete with you.

Note: I've attached a template NDA at the end of the book. Use it if you want, but don't engage in any real "negotiation" of the points in there without consulting an attorney. It is a very standard, down-the-middle NDA so anyone who objects to signing it probably should also have to carry a red flag around town because they're already sketchy.

Patents vs. Trade Secrets: Which Is Better for Creators?

This depends on what you're trying to protect.

- **Patents.** Best for something you want to share publicly but also own exclusively.

  - *Example:* A beauty influencer patents a new design for a ring light that attaches to smartphones and makes large noses look slightly less large. She will have to disclose everything about how to make it, how it works, the technical specifications, and the tips and tricks to using it effectively. If it has never been done, she will get the patent and can keep others from making an identical, cheaper version in China and competing with her. That said, if she doesn't get the patent, she's just told everyone how to compete with her – in exquisite detail.

- **Trade Secrets.** Best for something you want to keep entirely private.

  - *Example:* A gaming creator's proprietary strategy for maximizing ad revenue remains a secret known only to their close team. That close team, of course, should have signed an ironclad NDA and, if possible, more complicated agreements that can also be enforced if they skip town and open a competing brand in the next state.

Keep in mind that trade secrets last as long as you keep them secret, while patents eventually expire (usually after about 15-20 years).

Real-World Possibilities

1. **The Viral Algorithm Whisperer.** A TikTok creator developed a method for consistently landing on the For You page. By keeping the details private and only sharing them with close collaborators under NDAs, they've monetized their expertise through exclusive consulting gigs.

2. **The Innovative Podcaster.** A podcaster invented a unique soundproofing panel specifically designed for small recording spaces. She secured a utility patent, allowing her to license the design to other podcasters and users.

3. **The Merch Maven.** A lifestyle influencer created a line of eco-friendly packaging for her merchandise. She chose to patent the design to prevent competitors from copying it, securing their niche in the market.

The reality is that these types of situations are rare and not likely to come up very often, especially with patents. Trade secrets, on the other hand, may be more likely to come into play. The take-away should be in finding a comfort level to be able to identify trade secrets, particularly if you find they are generating real results. Don't go telling the world about it until you speak with an attorney. Once a secret is out of the bottle, there's no putting it back.

Practical Tips for Creators

1. **Evaluate Your Innovations.** Think critically about what you've created. Is it truly unique enough to patent, or is it better kept as a trade secret?

2. **Take Legal Steps Early.** Whether it's applying for a patent or drafting NDAs, don't wait until someone copies you to act. Be proactive, not reactive.

3. **Invest in Expertise.** Consult with an attorney to explore your options. Even a one-time consultation can clarify a lot.

4. **Stay Informed.** Laws around patents and trade secrets vary by country. Make sure you're aware of the rules where you operate. A U.S. patent will not protect you in China. If you want to enforce against a Chinese company violating your patent and selling to the U.S., I wish you the best of luck. In these sorts of cases, don't call me because I have better things to do than run my bald head into brick walls for the next 9 months. Sad fact of life, I'm afraid.

By understanding patents and trade secrets, you can protect your innovations and creative edge. These are specialty areas of law, and you should always seek out experts in these areas, particularly patent law. If you decide to patent something, start early and be prepared to wait. If you get a patent, don't sleep on it – enforce it! If you have a trade secret and it gives you an immediate advantage in the marketplace, don't share it, and start using it to your advantage. May be worth a lot more than a patent when all of the dust settles.

In the next chapter, we'll dive a bit deeper into the murky waters of copyright and trademark infringement —what it looks like, how to avoid it, and what to do if someone targets you with a claim.

CHAPTER 5

Defending Common Claims & Disputes

More than anything else in the digital world, you're going to be dealing with copyright claims and disputes. They are like the potholes on the road to success for content creators. You might be cruising along, building your brand, and then—BAM!—you hit a copyright issue that disrupts everything. Doesn't mean you won't be getting back on the road, but it will be a pain in the ass for a bit. In this chapter, we'll focus on things you can (but don't have to) do on your own to address common issues as they arise (e.g., takedown notices, copyright claims, etc.). Most of these things are done with little or no "legal" intervention and largely take place within the platform ecosystem.

We'll explore some of the usual copyright claims, how to avoid them, and how to (wisely) deal with disputes when they arise. This chapter has a more defensive focus (i.e., what happens when someone comes after you). In the next chapter, we'll focus on enforcing your rights (i.e., going on the offensive).

Both involve business decisions that you will need to make, some more serious than others. They can run the gamut between just deciding to take down a post to engaging an entire legal team to take down an infringing party in court. Once you know the landscape, I'm confident you can handle the majority of issues on your own. The key to being a smart steward of your own content is knowing if and when to escalate the issue to "legal."

Pro Tip: It's usually a lot sooner than you think. But, no worries. Lawyers are used to putting out fires after they are already burning wildly out of control. Check out Chapter 13 which covers some of the nastier disputes and, yes, by "nasty," I mean expensive.

What Is a Copyright Claim?

A copyright claim arises when someone believes that you have used their copyrighted material without permission. This might include:

- Using copyrighted music in your videos.

- Reposting someone else's artwork without credit or permission.

- Including clips or images from movies, TV shows, or other media.

How Claims Work on Major Platforms

Platforms like YouTube, Instagram, and TikTok have systems in place to detect and handle copyright issues (often called "strikes"). These systems can:

- **Automatically flag content:** Content ID on YouTube scans uploaded videos for matches with copyrighted material.

- **Issue takedowns:** Platforms may remove your content without notice or complaint from the copyright holder.

- **Demonetize your video:** On YouTube, for example, the revenue might even be redirected to the copyright holder.

- **Terminate your account:** Getting multiple strikes could result in your account being suspended or terminated.

Example: A YouTube creator uploads a vlog with a popular song playing in the background. Content ID flags the music, and the ad revenue from the video is automatically sent to the music's copyright owner.

There is a huge presumption, of course, that

YouTube has acted appropriately by redirecting what should be your money to the account of another or even to withhold it from you. As you recall, "fair use" is a permissible defense; apparently not one that YouTube is willing to engage in right off the bat. This puts you in the position of having to spend money on lawyers to try to get access to money that belongs to you. On a post-by-post basis, this might not be worth the effort, but if they demonetize you completely or shut down your account, you may have no choice but to "lawyer up."

NERD ALERT: There are a number of policy issues at stake with platforms like YouTube and Instagram inserting themselves into copyright claims without anyone asking them. While they might argue that they might face exposure from copyright holders if they didn't do it that way, this argument is mostly specious. They face exposure from one side no matter what. They are simply indicating that, as between individual content creators and well-heeled movie studios and record labels, they are going to err on the side of the big guy against the little guy. Hardly shocking, though they seem to be a lot more sanctimonious about it than they need to be. With the advent of the DCMA, they are now suggesting this is all to avoid losing the "safe harbor" protections of that law. Maybe. It's too deep a dive to get into here, but it's interesting if you have a few hours to spare and/or a strong desire to find something to put you to sleep within minutes.

Avoiding Copyright Claims

The best way to deal with copyright claims is to avoid them altogether. Here are some ways to do that:

1. Use Licensed or Copyright-Free Content

- Only use royalty-free music and images from reputable sources like Epidemic Sound, Artlist, or Unsplash. Make sure to document your licenses and sources to avoid a huge problem down the road.

- *Example:* A gaming streamer uses a track from a licensed music library in their streams to avoid issues.

2. Understand Fair Use

- Fair use allows limited use of copyrighted material without permission, but it can be context-specific inquiry and, unfortunately, often a legal gray area. Remember, courts consider factors like:

  - *Purpose*: Is the use transformative (e.g., a parody or critique)? Funny is better. Criticism is okay, but can easily devolve

into just watching the copyrighted piece
– that's not transformative or creative.

o   *Amount*:   Are you using only a small
portion of the work? Less is more. No
spoilers.

o   *Effect*: Does your use impact the market
for the original? Are people coming to
your site instead of theirs? If so, that's
bad.

• *Example:* A movie reviewer includes
short clips to illustrate their points.
While this might qualify as fair use, it's
not guaranteed.

3. Create Original Content

• The simplest (and, at the same time, most
difficult) way to avoid claims is to make
everything yourself. This means recording
your own music, taking your own photos,
and so on. Even as I write this, I know it
sounds ridiculous ... *"If I could record my own
music, I wouldn't need to use Led Zeppelin, now
would I? Thanks for nothing, Captain Obvious."*

4. Get Permissions or Licenses

- If you want to use copyrighted material, reach out to the copyright holder for permission or purchase a license. There are tons of services that license music; you don't have to track down Bruno Mars or Beyonce and get them to sign a license agreement. Also, pay attention to the platform's terms of use and limitations on length. The more you can avoid their copyright algorithm by staying within the guardrails, the better you will be.

- *Examples:* Here are some of big players in the music licensing business: ASCAP, BMI, SESAC, Global Music Rights (GMR), Re:Sound, SOCAN, Soundstripe, Marmoset, Pond5, Premium Beat, etc.

  o *Platforms (e.g., TikTok, Instagram) also have their own (perhaps limited) libraries of usable and licensable music. They can refer you to preferred third-party vendors to help avoid copyright issues, so it's worth doing some research on this front, as well, if you use a lot of music.*

When You Receive a Copyright Claim

Despite your best efforts, you might still receive

a claim. Here's how to handle it:

1. Don't Panic

- Claims are common and don't always mean you're in legal trouble. Many are automated and can be resolved easily.

2. Review the Claim

- Check the details:

  o Who filed the claim?

  o What content is flagged?

  o Are you at fault? Be honest with yourself. It will make it easier later.

3. Decide on a Course of Action

- **Accept the Claim:** If the claim is valid and minor (e.g., a snippet of music), you might choose to let it stand.

- **Dispute the Claim:** If you believe the claim is false or you qualify for fair use, you can contest it.

- **Remove the Content:** If the claim is serious, consider editing or taking down the content.

It's better to take it down, even if you think you're right. Damages will be calculated on the basis of continuing use of the infringing content, so consider that when you're deciding whether or not to pull it from your site.

Examples of Copyright Disputes

1. False Claims

False claims are a growing problem, particularly with automated systems.

- *Example:* A growing problem exists within the classical music industry where performers who upload their rendition of pieces in the public domain get regularly bombarded by copyright claims from music publishers. These music companies have automated bots that scan videos for notes similar to pieces they have under license. In December 2023, a 13 year-old reported that she had received over 100 copyright claims about her versions of public domain songs played on her violin. She reported having had videos removed or her performances muted by platforms based on the (false) determination of the bots alone. A number of other classical musicians have had similar experiences and have suggested it's a huge and expanding problem. Very often, the disputes can take months to clear up and, oh by the way, the platforms often defer to the bots that got the claim wrong in the first place. Check out the link below for the story: https://www.violinist.com/blog/laurie/202 312/29841/

2. Fair Use Battles

- Creators who believe their use qualifies as "fair use" may still face uphill battles.

- *Example:* A streamer uses a game's trailer in their review. The game's publisher files a takedown notice, despite the use being arguably transformative. It may even be complimentary. Doesn't really matter. Fighting that battle may be pointless. Stated differently, the juice may not be worth the squeeze.

3. Overreach by Copyright or Brand Owners

Sometimes copyright holders or brands file claims simply to silence criticism or even to stifle competition.

- *Example:* A beauty vlogger reviews a product critically, including clips from the brand's ads. The brand files a copyright claim to suppress the review.

Pro Tips for Handling Disputes

1. Document Everything

- Keep records of licenses, permissions, and any correspondence with copyright holders.

- *Example:* A creator who licensed a song provides proof when a claim is made. This is your best defense, so make sure to keep good records.

2. Be Professional in Disputes

- Avoid emotional or combative language. Stick to the facts and explain your case clearly.

3. Know When to Escalate

- If a platform's internal process doesn't resolve the issue, you may need to consult an attorney or take legal action. They get thousands of issues a day, and don't expect them to think your problem is as critical as you might think it is.

4. Consider the Risks

- Pursuing a claim in court can be costly and time-consuming. Evaluate whether it's worth the effort. As you'll see in Chapter 13, litigation is ungodly expensive and a time-consuming process. Here's a rule of thumb, though: however much you think it is going to cost (or think it should cost), double it and, then, double it again. Time is money, and these battles can take a lot of time. If you're still interested in getting in front of a judge or jury, let's go. I'll dust off my bow ties and get ready to throw down. I'm always up for that.

The Human Side of Claims and Disputes

Copyright disputes can be frustrating and even demoralizing. For creators, your work is personal, and a claim can feel like a direct attack. Just remember:

- These systems are far from perfect, and mistakes happen.

- Focus on creating and learning rather than getting bogged down in disputes.

- Every creator faces challenges; what matters is how you respond to them.

Building a Copyright-Proof Strategy

1. Educate Yourself

Take time to understand copyright laws and platform policies. The more you know, the better you can protect yourself.

2. Plan Ahead

Before posting content, consider:

- Have you checked for potential copyright issues?

- Do you have the rights to everything you're using?

3. Use Tools to Your Advantage

Platforms often provide tools for managing copyright claims. Learn how to use them effectively.

- *Example:* YouTube's Content ID lets you see claims and monetize disputed content while the issue is resolved.

It's All Part of the Game

Copyright claims and disputes are a reality for content creators. Part of the game, really. By understanding how they work, taking steps to avoid them, and handling disputes professionally, you can keep your creative journey on track and minimize disruptions. The consequences can be severe (e.g., account termination) so it's worth paying some attention to learning the rules of the game.

In the next chapter, we'll explore how to enforce your rights and go on the offensive, if it comes to that.

CHAPTER 6

Enforcing Your IP Rights

You've worked hard to create something original, protected it under intellectual property (IP) laws, and built a brand or business around it. But what happens when someone decides to take your work, profit from it, and/or damage your reputation in the process? This chapter will guide you through enforcing your IP rights, from recognizing infringement to taking legal action when necessary. Time to start pushing back.

LFG!

What Does IP Infringement Look Like?

IP infringement occurs when someone uses your protected work without your permission. Here are common examples for content creators:

- **Copyright Infringement**: A company uses your video or artwork in their marketing campaign without credit or payment – or permission. It's yours and you decide how it is going to be used. If you are irate about a particular political candidate using your copyrighted material, neither attribution nor payment may change your mind. It's not up to the candidate; it's up to you. But, usually, they have acquired a license from a third party (like BMI or ASCAP) and, in such cases, there is no infringement. Take to X and complain, by all means.

- **Trademark Infringement**: Someone starts selling merchandise using your logo or brand name. "Infringement" is one word for it. "Theft" is really more appropriate.

- **Patent Infringement**: Another business copies your innovative product or technology.

- **Trade Secret Misappropriation**: A former collaborator discloses your confidential strategy or technique to competitors or uses it to undermine your competitive advantage in a competing business.

How to Spot Infringement

Infringement can sometimes be obvious, but it's

not always easy to detect. You need to stay vigilant:

- **Set Up Alerts**. Use tools like Google Alerts to track mentions of your brand or content.

- **Use Reverse Image Search**. Platforms like TinEye or Google Images can help you find where your visuals are being used online.

- **Monitor Social Media**. Search hashtags, posts, and accounts for unauthorized uses of your work.

- **Watch Competitors**. Keep an eye on businesses or creators in your niche who might be tempted to borrow from your ideas.

There's an old saying that "imitation is the sincerest form of flattery." While that may be true when it comes to parenting, it's legally actionable if you have protected your rights and someone is taking food out of your kids' mouth. Look, criminals are generally lazy. They will rip things off rather than exercise their own brains to create original works if they believe they can get away with it. They may even do it, knowing they won't get away with it (i.e., just to piss you off). Don't try to understand the motivations of criminals. It's not worth it, and it's always the same reason anyway: laziness.

Responding to Infringement

When you discover someone is using your IP without permission, here's how to approach the situation:

Step 1: Confirm the Infringement

Before acting, make sure the use violates your rights:

- **Does it involve your protected IP?** Double-check that you have a valid copyright, trademark, patent, or trade secret.

- **Is it fair use or a valid exception?** For example, a brief clip of your video used in a critique may not qualify as infringement. Try to be objective about it. Don't just assume it's not a fair use and get all charged up. Look at the rules again (ahem, Chapter 2).

Example: A blogger reposts a photo you took on his Instagram page without credit. This violates your copyright unless they're using it under fair use (e.g., for commentary). How are they using it? Is it damaging or just annoying? Does it undermine your use of the photo? It's important to run the analysis before you get over your skis and start spending money on lawyers.

Step 2: Contact the Infringer Directly

Often, a polite but firm message can resolve the issue without escalating:

- Explain your ownership of the IP.

- Point out the unauthorized use.

- Request that they stop using the material, direct them to a place they can license the material, and/or provide appropriate attribution and compensation.

Example Template:

> *Dear [Infringer],*
>
> *I noticed that you are using my [specific work] without permission. As the copyright owner, I kindly ask that you [remove the material/credit my work/license it/compensate for its use]. Please respond by [specific date] to confirm that this issue has been resolved.*

Step 3: Use Platform Reporting Tools

Most major platforms have built-in tools for addressing IP violations:

- **YouTube**: Submit a copyright takedown request through the Content ID system.

- **Instagram**: Use the platform's copyright or trademark violation form.

- **Etsy or Amazon**: Report sellers who misuse your copyrighted designs or trademarks.

Example: A creator finds their artwork on unauthorized merchandise on Etsy. They file a takedown request, and the listing is removed within a few days. I would also recommend reaching out to the infringing party, as well, if only to let them know you are paying attention and you'll be watching them in the future.

Step 4: Send a Cease-and-Desist Letter

If the informal approach doesn't work, a formal cease-and-desist letter can escalate the matter. This letter should:

- Clearly identify the infringement.

- Demand that the infringer stop using your IP immediately or by a date certain.

- Warn of potential legal action if they don't comply.

While you can write this letter yourself, having it sent by an attorney often carries more weight.

Example: A photographer's image is used in a magazine without permission. Their attorney sends a cease-and-desist letter demanding removal and compensation, leading to a settlement.

The situation in the example does happen, and sometimes a strongly worded legal letter by itself gets the job done. But, be prepared to follow through on your threat of "potential legal action" if they just place the letter in their "circular file" with all the others. Threats of litigation are received differently by different people. Personally, if you wanted to sue me, I would expect you would just do it - not write a letter about *maybe* doing it. Overall, I'm not a huge fan of legal letters, but I understand that it might be a more cost-efficient solution for some folks. My experience has been that legal letters often don't work and that no one gets serious about settlement until a lawsuit is filed.

Taking Legal Action

If other measures fail, you may need to pursue legal remedies. While some of this is discussed in more detail in Chapter 13, here are your basic options:

1. File a Lawsuit

- **Copyright Infringement**: Seek damages for unauthorized use of your work.

- **Trademark Infringement**: Stop someone from using your brand and potentially recover profits they earned.

- **Patent Infringement**: Prevent further use of your invention and claim damages.

- **Trade Secret Misappropriation**: Recover losses (i.e., money damages) and stop the spread of confidential information (i.e., injunctive relief).

2. Seek an Injunction

An injunction is a court order requiring the infringer to stop using your IP. Failure to stop results in contempt of court and can get people arrested. This can be particularly useful for ongoing issues while a case plays out. Fair warning, though. Injunctions are not easy to get and can be document-intensive and, therefore, pricey. You may not have a choice, especially when trade secrets are involved.

Courts generally don't like to issue injunctions and prefer to just wait until the case is over and reduce everything to a monetary amount. Almost anything can be reduced to a dollar value, by the way. I could go into a lot of details as to why courts like to avoid injunctions, but as someone who has spent the better part of his professional life specializing in seeking and securing injunctions in state and federal courts, let me just say this:

If it can be reduced to money damages (i.e., if money will make you whole), you're not going to get an injunction. Trade secret cases have better odds, injunction-wise, but you better have persuasive evidence that you have taken great pains to keep that information confidential and how it is a genuine, valuable "secret sauce" that you created and did not intend to disclose.

3. Pursue Damages

Courts may award damages for:

- Lost profits.

- Royalties you would have earned.

- Statutory damages for registered copyrights or trademarks (sometimes legal fees are recoverable, as well).

Example: A content creator discovers that her viral video has been used in a documentary without consent. After filing a lawsuit, the court awards her damages for both lost revenue and emotional distress (let's assume the video was less than flattering). Damages can be compensatory (i.e., designed to compensate you for your loss) and, in some cases, punitive (i.e., designed to punish the wrongdoer). Usually, you're only going to get the former.

Common Challenges in IP Enforcement

Enforcing IP rights can be tricky. Here's how to navigate common hurdles:

1. Jurisdiction Issues

Infringement by someone in another country can complicate enforcement. You may need to hire legal experts familiar with international IP law. You may have to hire counsel in another country. You may just get hosed even if you do all of that. I represented a client in a trademark fight in Mexico. We hired Mexican IP counsel, who demonstrated proof of the other company ripping off my client. The Mexican court ignored it and sided with the home company without so much as a reason given. If it's China, forget about it. It will be nearly impossible to win. They've taken IP infringement to new heights over there, almost as though it was a sovereign policy...

2. Cost of Legal Action

Again, lawsuits can be expensive. Assess whether the potential recovery justifies the cost. I've addressed this already (recall the "double then double again" rule of thumb). Ask yourself this question, too: What does a "win" look like? Then, compare that to the pile of money it's going to cost to get it. Decide which is more important to you right now.

3. Difficulty Proving Infringement

Gathering evidence is critical. Keeping comprehensive and organized records is invaluable. That alone can save tens of thousands of dollars in legal costs down the road. Save screenshots, URLs, and other proof to strengthen your case.

Proactive IP Enforcement Strategies

Rather than waiting for infringement to happen, take steps to protect your work proactively:

- **Watermark Your Content**: Add subtle watermarks to images or videos to deter misuse.

- **Use Licensing Agreements**: Clearly outline how others can use your work and under what conditions.

- **Register Your IP**: While copyright is automatic, registration strengthens your legal position (required if you want to sue).

- **Work with Monitoring Services**: Services like Pixsy and BrandShield can help track and enforce your IP rights. There are vendors that work with law firms, as well, that can conduct valuable searches and scans of potentially infringing content.

Example: A digital artist includes a watermark on her portfolio images, deterring unauthorized use and encouraging clients to purchase licenses instead. You've certainly seen watermarks on images from the big stock photo sites (e.g., Getty Images, Adobe Stock).

We'll talk about it more when we discuss licensing, but be mindful of the terms of the license itself. Paying for the "personal use" license only to turn around and use it commercially will get flagged, and those marketplaces will come after you to make you pay for the right license. It may be years later, by the way.

Real-World Case Studies

Case 1: Taylor Swift's Trademark Savvy

The pop star trademarked phrases from her lyrics, like "This Sick Beat," to prevent others from profiting off her brand. She started early in her career (2008) and has not let up, domestically or abroad. This proactive approach showcases the importance of trademarking valuable phrases. When you consider live and pending applications worldwide, Tay-Tay's number approaches 500. Even her cats' names are trademarked! That's what you call a trademark strategy.

Case 2: Disney's IP Enforcement Machine

Disney is notorious for aggressively protecting its IP, even targeting small businesses using unlicensed Mickey Mouse imagery. In 1989, Disney demanded three Florida daycare centers remove life-sized murals of characters Disney's attorneys determined were infringing Disney's IP, the disappointment and tears of the children at those facilities notwithstanding. While their approach can seem excessive, it underscores their "zero tolerance" approach. This, by itself, has a deterrent effect on future infringers. Recall, though, that some of that Disney IP has hit the public domain. Disney may have to learn to live in a world where its iron-handed approach bears less fruit than it used to. Maybe a few daycare centers in Florida might want to break out the paint brushes again...

Disney is certainly not alone in using this approach. Others include Apple, Nike, Louis Vuitton, and Jack Daniel's. The whiskey maker has been known to be a bit more playful with their enforcement, coming up with creative settlements to avoid potential disputes and earning the brand some acclaim in the process. They take IP protection seriously, but they don't take themselves too seriously, it seems. Disney might want to take a few notes.

Case 3: A Content Creator's Viral Photo

A travel photographer's photo went viral,

appearing on dozens of websites without permission. Using reverse image search and sending DMCA takedown requests, they successfully removed most unauthorized uses and secured licensing fees from others.

Let's Talk about Your Nudes...

You know the ones. The ones you thought nobody knew about, hidden away on your computer in folders marked "Taxes." Yes, those. What happens when you wake up one morning and realize they have found their way to the internet (and they always do, by the way) without you knowing about it? Current employers, ex-husbands, judges, parents, neighbors, future employers, creepy uncles, busboys, pastors, baristas, and bank personnel seeing you in your birthday suit can (and will) be a humbling experience, to put it mildly. Leaked Only Fans images, shots you took with your ex-boyfriend in Cancun, videos taken by hidden camera in that sketchy AirBnB, and secret photos you only shared with your BFF who now is dating your ex-boyfriend – all of these could see the light of day whether you want them to or not.

Enough talk. How do we get them taken down?

If you were paying attention above, your naked

image, in and of itself, is not generally copyrightable. It's not infringement of your copyright if someone takes a naked picture of you. In fact, they own the copyright on that photo automatically. If you take a picture of yourself in the mirror, though, that's your copyrighted property: the photo, not the "image" in the photo. There may be a handful of copyright issues in play (e.g., leaked Only Fans images), but mostly you're going to be dealing with what are called privacy and publicity rights (which vary by state), especially when you're being filmed without you knowing about it. Some states have also passed "revenge porn" laws that come into play in certain situations and can be helpful.

If we had to come up with a playbook to get your behind off of the front page of the internet, let's focus on doing the following:

1. Document the unauthorized use – take screenshots and get dates/times/etc. Try reverse image searches (Google Images, Bing Image Match) to track down where the image has been posted. There are other specialty firms that can help, too (Pixsy, TinEye, IMATAG Monitor).

2. Contact the website and demand the removal of the images.

3. If it is copyrighted material (e.g., a mirror selfie you took in your bathroom), issue a DMCA takedown request to the website. If it's not

copyrighted material (e.g., a picture your ex-boyfriend took on his phone), send a cease-and-desist letter and a DMCA takedown request (couldn't hurt – you can argue about who owns the copyright later).

4. Utilize Google's removal request process to bring down the images – which must be explicit or intimate *and* used without your consent. Go to Google Search Help to begin that process. There are other search engines and they have similar processes in place to help. Don't just stop at Google.

5. Report the person who posted the image to the social media platform (as applicable), citing their community guidelines. Again, there are a bunch of them, so you'll need to do this a few times, perhaps.

6. If these things don't work (or only partially work), contact a lawyer ASAP. Time is not your friend, here. Ideally, you should have a lawyer do all of the steps above, but it's your call. Your lawyer may need to file a lawsuit for defamation, invasion of privacy, infliction of emotional distress, misappropriation of likeness, and statutory violations – the actual claims will vary from state-to-state – in order to get them taken down.

Pro Tip: Just because a website removed your naughty image, that doesn't mean it won't still show up in search results. You may need to submit separate removal requests to multiple search engines after it is removed from the original site.

Be careful out there.

The Best Defense is a Good Offense

Enforcing your IP rights isn't just about protecting your income - it's about safeguarding the hard work and creativity that make your content unique. Pirates don't fly the Jolly Roger anymore. They are real, and they are out there. As someone who spends the better part of his life in the middle of disputes, my advice is simple: bullies don't respond to legal letters. You need to punch them in the mouth. Legally-speaking, of course.

By staying vigilant, taking proactive measures, keeping good records, and knowing when to escalate disputes, you can ensure that your IP remains yours.

In the next chapter, we'll examine how IP issues come up when dealing with a number of the most popular digital platforms.

CHAPTER 7

Navigating Content Platforms and Their Policies

Why Platform Policies Matter

For content creators, platforms like YouTube, Instagram, TikTok, and Twitch are the battlegrounds where your brand is built. These platforms provide an audience, monetization opportunities, and tools to amplify your work. But they come with rules—and ignoring those rules can lead to takedowns, demonetization, and even bans.

Here's the truth: every piece of content you upload must comply with the platform's terms of service (TOS). I know you didn't read them. I know you just checked "Accept." We've all done it. No judgment. But, those things are legally binding agreements that govern what you can and cannot do on the platform. Violating them can be a costly mistake, so understanding the nuances is critical.

Key Elements of Platform Terms of Service

Let's break down the most important aspects of TOS you need to understand:

1. **Content Ownership**

   - Most platforms state that you retain ownership of your content.

   - However, by uploading, you often grant the platform (or the owner of the platform – looking at you, Facebook) a broad license to use, display, and distribute your work. They don't necessarily have to tell you about it, and they don't have to pay you for it.

   - *Example*: YouTube's TOS gives the platform the right to reproduce and display your videos globally to operate their services.

2. **Prohibited Content**

   - Each platform has strict rules about what's not allowed. This often includes hate speech, explicit content, or anything that violates copyright law.

   - *Example*: The Community Guidelines on Instagram prohibit sharing content that promotes violence or self-harm – in addition to prohibitions on the promotion of terrorism, illegal activities, organized crime, nudity, bullying, misinformation, buying/selling of firearms and tobacco, and impersonating another person. Are they violated every single day? Yep. Are they still violations if you get caught doing any of them? Yep.

3. **Copyright Enforcement**

- Platforms are obligated to comply with copyright laws like the DMCA (Digital Millennium Copyright Act). They often use automated systems to flag or remove infringing content.

- *Example*: A Twitch streamer's broadcast could be muted or taken down if copyrighted music is played during their stream. But, if you really want to get kicked off quickly, take your clothes off. That will get you banned from Twitch faster than you can, well, twitch. If you want a fun read, check out the Twitch Community Guidelines and, specifically, the Attire policy and the "contextual exceptions." It's a gem of a read.

4. **Monetization Rules**

- To earn revenue, you must follow platform-specific monetization policies. Violations can lead to demonetization.

- *Example*: YouTube's Partner Program requires adherence to advertiser-friendly guidelines. It's not always easy to know how or why you may have violated some of these guidelines. Things like adult content, profanity, and depictions of violence are not the kinds of things Proctor+Gamble want to advertise shampoo against as a general rule.

5. **Termination Rights:**

- Platforms reserve the right to suspend or terminate your account for violations, often without prior notice.

- *Example*: TikTok might ban an account for repeated policy violations, such as sharing misleading information. There's a fair measure of irony in TikTok banning people for misinformation, particularly given the lack of transparency that platform has demonstrated in terms of what it does with users' data. So much so that, in April 2024, President Joe Biden signed a law requiring TikTok's parent company (ByteDance) to divest itself of its ownership or face a nationwide ban (which did briefly happen in

January 2025). There were some over-dramatic tantrums and tears, of course.

How Platforms Enforce Policies

Understanding enforcement mechanisms can help you avoid problems and nasty consequences. Here are some common methods:

1. **Automated Content Moderation**

   - Platforms use AI to detect policy violations. This sort of tool is fast but not foolproof and can lead to false flags.

   - *Example*: Instagram might mistakenly flag an artistic nude as explicit content, even if it complies with their guidelines.

2. **User Reporting**

   - Users can report content they believe violates policies. A single viral complaint can trigger an investigation. This can, of course, be abused by competitors, jerks, and former BFFs.

3. **Copyright "Strikes"**

- For platforms like YouTube, repeated copyright violations (or, "strikes") can result in account suspension.

- *Example*: A gaming creator received 3 copyright strikes for using unlicensed music and lost their channel. No more new lives on that channel...

4. **Monetization Penalties**

- Violating advertiser-friendly guidelines can result in reduced (or no) revenue.

- *Example*: A TikToker's earnings dropped after being flagged for "controversial" content.

5. **Permanent Bans**

- Serious or repeated violations can lead to account termination, with little recourse.

- *Example*: A high-profile Twitch streamer was banned permanently for unapproved gambling content.

Remember, these platforms are privately owned. They are under no obligation to let you violate their rules and give you an outlet to encourage others to do the same. It is no defense to say that you're just doing what others are doing, either. If you don't like it, go hang out on MySpace.

Tips for Staying Compliant

Navigating platform policies can feel like walking a tightrope. Here's how to stay on the (relatively) safe side:

1. **Read the TOS and Guidelines**

   - Skim the key sections related to prohibited content, copyright, and monetization. Many platforms provide summaries or FAQs to make this easier.

2. **Use Royalty-Free Resources**

   - Avoid copyright issues by using licensed or royalty-free music, images, and video clips. Tools like Epidemic Sound or Canva can help.

3. **Avoid Reposting Without Permission**

   - Sharing someone else's content without credit or permission is risky, even if it's for a harmless meme.

4. **Flag Issues Proactively**

- If you think your content might trigger a false flag, consider explaining its context in the description or tags.

5. **Monitor Updates**

- Platforms frequently update their policies. Stay informed through their newsletters or help centers.

Examples of Policy Missteps

Sometimes creators unintentionally run afoul of platform rules. Here are some real-world scenarios:

1. **The Music Licensing Trap:**

- A YouTuber's entire video library was demonetized because they used popular songs without proper licensing. They had to replace all of the music and reapply for monetization.

2. **Community Guideline Violations:**

- A TikToker was temporarily banned for sharing edgy humor that allegedly crossed the line into "hate speech." The creator appealed and regained access but lost significant momentum.

3. **Platform-Specific Rules:**

- An Instagram influencer had a post removed for "promoting gambling" because it included a casino backdrop, even though gambling wasn't the focus.

When Things Go Wrong: Handling Takedowns and Strikes

If you're flagged for a policy violation, don't panic and, more importantly, don't crawl into the fetal position in the corner. Your life is not over. Here's what to do:

1. **Review the Notice:**

- Platforms usually provide a reason for the action. Read it carefully to understand the issue.

2. **File an Appeal:**

- Most platforms allow appeals for strikes or removals. Be clear and concise in explaining why you believe your content complies with their policies. Resort to pointing out other peoples' violations is tempting, but it doesn't make your case – it only points out flaws in their system. You're trying to get these folks on your side, remember?

3. **Get Advice:**

- If you're facing repeated issues, a lawyer experienced in digital content can help craft a stronger appeal or take further action. You might simply be too close to the issue to be a persuasive advocate for your position.

- If you have friends who have successfully navigated these issues, ask them for advice.

4. **Diversify Your Platforms:**

- Relying on one platform is risky.

- Building a presence on multiple platforms ensures you're not completely sidelined if one account is restricted. It's not unusual to see people using multiple accounts on the same platform – though, it should be noted, that looks a bit suspect (or, as the kids say, "sus") and will probably be viewed more as an admission that you knew your content violated the TOS long before they banned it.

It's Their World. You Just Live In It.

Platforms are the gatekeepers of your content's visibility and monetization. By understanding their policies and staying compliant, you can safeguard your account, maintain your income streams, and focus on what you do best – annoying your parents, dancing like an idiot, unpackaging boxes, crushing weird things with heavier things, pranking your friends, or annoying unsuspecting Karens in the wild by filming them. Whatever it might be.

Just remember, these platforms consider you to be a guest in their house (more like a really expensive AirBnB, if we're honest). If you abuse the house rules (no matter what you might have paid), you will be on the outside looking in, asking strangers for nickels on the sidewalk.

In the next chapter, we'll go a bit deeper into issues surrounding monetizing your content while protecting your legal rights.

CHAPTER 8

Monetizing Your Content While Staying Legal

The Legal Side of Turning Creativity Into Cash

For content creators, monetization isn't just about creativity; it's also about maximizing opportunities while staying within the bounds of the law. Whether you're selling merchandise, running ads, or securing brand deals, legal missteps can derail your income and damage your reputation (unless you're one of the Kardashians, it seems – more on that later).

In this chapter, we'll explore the best ways to monetize your work while protecting yourself from legal pitfalls. Those I can help you with.

That said, there's no way I can possibly protect your dignity or self-respect. That's on you, I'm afraid.

Popular Monetization Methods for Creators

Creators have more options than ever to generate revenue. In your mind, no matter the amount of followers you have, you should always be thinking about how to capitalize on that following and make money on it. Let's face it. The "success stories" and "icons" of the trade are not dealing in fine art. Often, they're dealing in shameless titillation and base-level voyeurism, FOMO joiners, and, yes, card-carrying mouth-breathers that drive dollars to your bank account, even though (and maybe especially if you think) you have real talent.

Ask the D'Amelio girls about the difference between being legitimately and objectively talented on the one hand and, on the other, being social media darlings who get paid to suggest the talent they do have is actually legitimate and objectively interesting. Boggles the mind, I realize. But, 100 million followers don't lie. Tens of millions of dollars in their bank accounts doesn't lie. Long term success? Maybe not. I just hope they have good financial managers (and that the sex tapes don't surface).

For the vast majority of influencers and creators, here are the most common avenues:

1. **Ad Revenue**

   - Platforms like YouTube, TikTok, and Instagram share ad revenue with eligible creators (i.e., those with a certain number of subscribers and/or hours of content watched) – provided, of course, you adhere to the advertising guidelines.

   - *Example*: A YouTuber with a monetized channel earns a fair portion of the revenue from ads shown before or during their videos. Yes, those ads are annoying. Yes, that "Skip" button is too small. But, they pay the bills and keep lights on. Some quick math: In 2023, YouTube generated roughly $31 billion in ad revenue, with roughly 55% of that going to creators. That's a little over $17 billion being handed out. Some might argue the percentage is a bit on the low end, but it's still a huge chunk of change that ends up in the pockets of people who film cats, power-washing, questionable recipe generation, and their idiot friends falling down stairs.

2. **Sponsorships and Brand Deals**

   - Companies pay creators to promote their products or services. Brands are attracted to creators with a massive following like moths to a new generational flame. Some of the time, it even works. (Cautionary tale: Dylan Mulvaney and Bud Light – did not work). Clever idea or terrible one, transparency is key to avoid running afoul of FTC rules around deceptive advertising and disclosure of paid content (more on that below – pay attention, Kim, Khloé, Kylie, and Kourtney).

- *Example*: An influencer collaborates with a skincare brand and discloses the partnership in her posts. You will often see the words "Sponsored Post" or #sponsored or #ad somewhere in the post. Get in the habit of not forgetting to do that, by the way.

3. **Merchandising**

- Selling branded merchandise like T-shirts, mugs, or digital products.

- *Example*: A Twitch streamer offers custom emojis and exclusive merch to their subscribers.

4. **Subscriptions and Memberships**

- Platforms like Patreon, Only Fans, and YouTube Memberships allow fans to pay for exclusive content or perks by their favorite personalities, often giving away only minimal content for free, with extended (sadly, too often explicit) content behind the "pay wall."

- *Example*: A podcaster offers behind-the-scenes episodes, bonus content, ad-free feeds, and other benefits to their paying subscribers. It has become an increasingly common phenomenon, especially with the crowded nature of the podcast space. Giving hardcore fans an inside look at the way their favorite podcasters make the sausage is the end-all-be-all of passive income. Adam Carolla, Joe Rogan, and a handful of others have given a masterclass on how to monetize the same content several different ways. Get it on!

5. **Affiliate Marketing**

- Promoting products and earning a commission for each sale made through your referral link.

- *Example*: A beauty blogger shares affiliate links for makeup products and earns a percentage of each sale. Some might offer promotional codes or ask you to access sites like Amazon through hyperlinks on their website or bio.

- *Example*: A podcaster reading a live advertisement for a sponsor during the show is worth even more. We get it, Ben Shapiro, you really like ExpressVPN.

Legal Considerations for Monetization

Each monetization method comes with its own legal obligations. Let's break them down:

1. **Ad Revenue**

   - Platforms often require adherence to advertiser-friendly content guidelines. Violating these rules can lead to demonetization.

   - *Key Tip*: Review the platform's monetization policies regularly (they can and do change). Content flagged as "sensitive" or "controversial" may lose ad revenue altogether or be significantly reduced.

2. **Sponsorship Disclosures**

   - Transparency is a legal requirement. The FTC (Federal Trade Commission) mandates that sponsored content must be clearly disclosed to viewers. See below for an extended look into the many and various violations of the FTC rules by the Kardashian clan.

- *Example*: Phrases like "Sponsored by [Brand]" or "This video contains paid promotion" are required to ensure compliance.

3. **Taxes**

- Revenue generated through ads, merchandise, or memberships is taxable income. Keep detailed records of your earnings and expenses.

- *Key Tip*: Consult a tax professional to ensure you're deducting eligible business expenses, such as equipment, licenses, subscriptions, and platform fees. You're a business owner. Think like one.

4. **Intellectual Property Issues**

- Ensure you have the right to use all assets in your monetized content, including music, images, and video clips.

- *Example*: A gaming YouTuber needs permission to monetize gameplay from certain publishers.

5. **Contracts for Brand Deals**

- Always review sponsorship agreements carefully. Look for clauses about deliverables, payment terms, contract length, and usage rights. We'll discuss contracts in more depth in Chapter 10.

- *Key Tip*: If a brand requests perpetual rights to your content, get real nervous. Negotiate for a (much) higher fee or, else, demand a time-limited license.

6. **Merchandise Compliance**

- Be mindful of copyright and trademark laws when designing merchandise. Using copyrighted images without permission can lead to legal trouble, especially if you don't want to give back all of the money you earned on the merch.

- *Example*: A creator selling T-shirts with characters from a popular TV show could face infringement claims. If that TV show happens to be owned by Disney, their army of lawyers might just bulldoze your house. Kidding (but only sort of).

Common Legal Pitfalls and How to Avoid Them

Even seasoned creators can stumble into legal issues. Here are some common mistakes:

1. **Ignoring Disclosure Rules**

   - Many creators fail to properly disclose sponsored content, leading to FTC warnings or fines.

   - *Solution*: Use clear and conspicuous language like "Ad" or "Paid Partnership" in your content. While hashtags help, you might not want to just rely on #sponsored or #ad. Generally speaking, use #ad, which the FTC deems "clear and unambiguous" instead of #sponsored which has been deemed insufficient, particularly where it's not super-clear. Phrases like "collab" or "affiliate" or simply tagging the brand are not sufficient. In any event, make them clearly visible to the consumer. If you hide them, it doesn't matter what disclosure you make. It will be a violation.

2. **Using Unlicensed Music**

- Background music can lead to copyright strikes if it's not properly licensed.

- *Solution*: Use royalty-free music libraries like Epidemic Sound or Artlist.

3. **Overlooking Platform-Specific Rules**

- Platforms have unique monetization policies that must be followed. These rules can change and, weirdly, the platform's enforcement can and does change. Sometimes, the enforcement is just erratic or inconsistent. Make sure you know the rules - even if they might not.

- *Example*: TikTok, X, and YouTube prohibit affiliate links to certain types of content – even though those products may be perfectly legal to buy, consume, and own. Don't like it? Make your own platform.

4. **Neglecting Contracts**

- Skipping contracts or relying on verbal agreements for brand deals can, and often does, lead to disputes. Get it in writing and make sure the key deal points are clear.

- *Solution*: Always insist on a written agreement, even for smaller deals.

5. **Failing to Separate Personal and Business Finances**

- Mixing personal and business expenses can complicate tax reporting. We will get into this in more detail in Chapter 13, but not paying close attention can also undermine any special legal protections you might have put in place (e.g., forming a limited liability company or corporation). It's called "co-mingling" of funds and it is a telltale sign that your company does not deserve the "limited liability" protection those company types should give you. Think of it this way: if your properly-run LLC gets sued and loses everything, you don't lose your house, car, clothes, and comic book collection. If you co-mingle your LLC funds with your personal funds, you will lose all that stuff. Talk to a tax or legal professional if you are at all confused about setting up your business correctly.

- *Solution*: At a very minimum, open a separate bank account for your content creation business. Pay yourself a salary (even a really good one) and then that business money becomes your personal money to do with what you wish. Don't say meme coin. I hear Pokemon cards appreciate like gold but without the hassle of all the crypto bro interaction. Or, just a nice mutual fund, maybe.

Examples of Monetization Success Stories

Let's look at how some creators have turned their content into thriving businesses:

1. **The Merch Powerhouse**

   - A YouTuber in the tech space (Mrwhosetheboss) launched a line of branded phone accessories (Phone Rebel). By focusing on high-quality, niche products, he built a booming business, leveraging his 8 million subscribers into a successful stream of income.

2. **Patreon for Artists**

   - An illustrator named Becca Hall used Patreon to offer exclusive tutorials and artwork. By engaging directly with fans of all levels of ability, she has generated a steady monthly income that could very well be the envy of illustrators everywhere.

3. **Affiliate Marketing Pros**

- There is an army of fitness influencers who have partnered with supplement and fitness equipment brands, earning commissions through affiliate links. Their authentic reviews can drive significant sales and, hence, commission income.

- Numerous marketing polls have demonstrated that consumers believe influencers considerably more than, say, "celebrity" endorsements. By a large amount. There is something to be said for the intimate relationship with influencers that they have worked hard to foster. While bad recommendations can burn that relationship quickly, focusing on honest, authentic commentary can be hugely marketable – and profitable.

It's Time to Address the Kardashians & the FTC

The Federal Trade Commission (FTC) has become increasingly vigilant in regulating paid partnerships and deceptive advertising claims. Both areas have serious implications for businesses and influencers alike, as failure to comply with FTC guidelines can lead to significant penalties. The importance of transparency in advertising has never been clearer, as the FTC seeks to protect consumers from misleading or deceptive practices.

Failure to Disclose Paid Partnerships. One of the most common issues in influencer marketing is the failure to disclose paid partnerships. When an influencer is paid to promote a product or service, it is crucial that this relationship is disclosed in a clear and conspicuous manner to avoid misleading the audience. The FTC requires that influencers clearly state when a post is sponsored or contains affiliate links. Failure to do so could be deemed "deceptive advertising," and both the influencer and the brand can face consequences, including:

- **Monetary Penalties**: The FTC can impose fines for non-compliance with disclosure rules.

- **Reputational Damage**: Brands and influencers caught in deceptive practices may lose consumer trust, which can result in long-term damage to their reputation.

- **Legal Action**: In severe cases, the FTC can take legal action, resulting in settlements or injunctions.

Deceptive Advertising. Equally important are deceptive advertising claims, where a business or influencer misrepresents the effectiveness, safety, or benefits of a product. The FTC has strict guidelines on what constitutes deceptive advertising, and advertisers are prohibited from making false or unsubstantiated claims about products.

These claims could be about a product's ability to cure diseases, improve health, or provide unrealistic financial benefits. Consequences for deceptive advertising include:

- **Injunctive Relief**: Companies may be required to stop making certain claims or even cease selling a product.

- **Corrective Advertising**: In some cases, businesses are ordered to issue corrective ads to clarify misleading information.

- **Lawsuits and Civil Penalties**: The FTC may sue for damages if a business violates its guidelines on deceptive advertising.

- **Other Potential Penalties**: Social media platforms like Instagram, TikTok, and YouTube have their own rules and community guidelines that influencers must follow. If an influencer repeatedly violates these platform policies, such as failing to disclose sponsored content, engaging in deceptive advertising, or breaching other rules (like promoting harmful products), they could face penalties ranging from temporary suspensions to permanent bans.

Lord & Taylor, Carrefour, Sony, Rosie Huntington-Whitely, Lauren Conrad, Michelle Lewin, Chloe Morello, and a host of others have all stared down the barrel of an FTC investigation into some form of deceptive advertising or disclosure failures. But, they are just amateurs, really.

When it comes to the Kardashians, the run-ins are so frequent and shameless that it almost seems like some sort of perverse "strategy." They are truly in a class of their own.

Various members (and hangers-on) of the Kardashian-Jenner clan have, at one time or another, had high-profile run-ins with the FTC over their social media and advertising activities, including:

- **Kim Kardashian**: In 2016, Kim Kardashian was criticized for not disclosing her paid partnership with the diet supplement brand, **Flat Tummy Co.**, when she posted about its products on Instagram. Although she later acknowledged the mistake, the incident highlighted the importance of clear and conspicuous disclosures. She was, along with other influencers, reminded by the FTC to clearly label sponsored content, especially when promoting products that make health claims.

- **Kendall Jenner**: Kendall Jenner faced similar issues with **Pepsi** in 2017, though it was not about paid partnerships. However, her endorsement of the brand in the controversial Pepsi ad led to public backlash, and the FTC investigated the ad's (arguably) misleading messaging. The incident served as a cautionary tale about the FTC's interest in deceptive advertising, particularly when it comes to using popular figures and influencers to promote a brand.

- **The Kardashian-Jenner Family**: The family as a whole has been repeatedly scrutinized for not properly disclosing their paid promotions on platforms like Instagram and Snapchat. For example, they were reportedly reprimanded after not including clear disclosures in posts about products like **Flat Tummy Tea** and other various beauty products. As a result, the FTC issued warnings to all influencers and brands about the need to clearly label sponsored posts to avoid misleading their audiences.

- **Scott Disick**: Another member of the Kardashian-Jenner extended family, Scott Disick, faced FTC scrutiny in 2018 for promoting a **Fit Tea** detox product without a proper disclosure. In response to this, the FTC issued a warning to both Disick and the company for failing to adhere to disclosure guidelines.

- **Kylie Jenner – Sponsored Post Issues with "Pouty Lips"**: In 2017, Kylie Jenner faced scrutiny for promoting her own lip kits and other beauty products on Instagram without sufficiently disclosing the nature of her paid partnership with her own company. Although the posts appeared promotional, they did not have clear labeling like #ad or #sponsored. The FTC later reminded influencers, including Jenner, of the importance of clear and conspicuous disclosures when promoting products, especially when the influencer owns the brand or stands to benefit financially.

- **Khloé Kardashian – FitTea Controversy**: In 2015, Khloé Kardashian was criticized for promoting **Fit Tea** detox products on Instagram without an adequate disclosure of the paid partnership. She posted pictures showing off the tea but did not include any prominent disclosure like #ad or #sponsored, which raised concerns about deceptive advertising. She later addressed the issue, but the FTC used the incident as an example to emphasize the importance of transparency in social media marketing. This says nothing about the fake accent mark in her name (Khloé), which, some might argue, is also deceptive!

- **Kourtney Kardashian – Poosh and Unclear Disclosures**: Kourtney Kardashian's lifestyle brand, **Poosh**, has also attracted FTC scrutiny due to unclear or insufficient disclosures in social media promotions. Some of her Instagram posts promoting products like health supplements and beauty items did not include prominent labeling that they were paid advertisements. The FTC has been particularly focused on influencers and celebrities who own or are closely involved with the brands they promote, as failing to disclose these relationships can mislead followers.

- **Kim Kardashian and Flat Tummy Co. (Again)**: In 2018, Kim Kardashian's continued partnership with the controversial weight loss brand **Flat Tummy Co.** led to multiple FTC warnings regarding her posts promoting the brand's detox teas and appetite suppressant lollipops. Despite being a paid endorsement, some of her posts did not use the required disclosure tags like #ad or #sponsored. This led to public backlash, and the FTC used her example to reiterate the importance of full and clear disclosure in all paid partnerships, especially when promoting health-related products.

- **Kim Kardashian's Instagram Contest – Lack of Disclosure**: In another FTC-related incident, Kim Kardashian was called out for promoting a contest on her Instagram page where participants could win a trip to Dubai. The post failed to properly disclose the nature of the contest, which involved a brand partnership. While this issue didn't result in an outright fine or penalty, it raised concerns about the lack of clarity in influencer-led promotions and contests.

The Kardashians are an object lesson in the old adage: "It's easier to beg for forgiveness than to ask for permission." While that strategy may (and seems to) work for the Kardashian ladies, you may not have the same luck.

All the more ironic, of course, is that each of the examples above came after a $40 million settlement in 2012 between the sneaker brand Skechers and the FTC involving deceptive claims in ads for their Shape-up line of shoes (e.g., weight loss, muscle strengthening, buttocks toning!). Can you name one of the stars of that ad campaign? Of course, it was Kim Kardashian.

Actionable Steps to Monetize Legally

Ready to take your monetization strategy to the next level? Follow these steps:

1. **Understand Your Platform**

   - Review the monetization policies of each platform you use.

2. **Get Organized**

   - Track your income, expenses, and contracts.

   - Use tools like QuickBooks or Wave to simplify bookkeeping.

3. **Build a Legal Toolkit**

   - Work with a lawyer to create templates for contracts and review brand deals.

4. **Invest in Licensing**

   - Use licensed music, images, and video clips to avoid copyright issues.

5. **Be Transparent**

   - Disclose all sponsorships and partnerships clearly.

6. **Get Accounting Help**

- Accountants are invaluable partners in any business. They are usually less expensive than lawyers, have a lot of clients who have seen a lot of things done to them by the taxing authorities, and can help you avoid a ton of easy mistakes. You don't want to be posting Instagram Reels with #irsaudit on them. Believe me, you do not want that.

Monetize Smarter, Not Riskier

Monetizing your content can be rewarding, sustainable, and downright lucrative when done right. Ask Jake and Logan Paul. Heck, even ask the Kardashians. While you may not be playing in their rarified air just yet, that doesn't mean you can't make a good bit of coin utilizing your own creativity. By understanding the legal landscape and taking proactive steps to protect yourself, you'll create a solid foundation for lasting success.

In the next chapter, we'll talk a bit about protecting yourself and your online presence, as well as your reputation.

CHAPTER 9

Protecting Your Online Presence & Reputation

Why Your Online Presence Matters

Your online presence is more than just a collection of posts, videos, or tweets; it's your personal or brand identity in a digital world that can be, frankly, brutal and cruel. A strong online reputation builds trust with your audience, opens up monetization opportunities, and sets you apart from competitors. However, the same digital landscape that allows you to thrive can also expose you to risks, including impersonation, defamation, and cyberattacks.

In this chapter, we'll address some practical steps for safeguarding your online presence and reputation, from managing your digital footprint to handling crises effectively. Every emergency is different, and every problem has a unique solution. How you respond is what makes you resilient and special. It is will what will carry you through the tough times. Fall apart, melt down, or get pushed around and you will fall flat. Be professional, stay confident, and make your case. You'll be fine.

The Importance of Proactive Reputation Management

Building and maintaining a positive online reputation requires continuous effort. Here's why it's crucial:

1. **Trust Equals Monetization**

   - Audiences are more likely to engage with creators and brands they trust, leading to better sponsorship deals, ad revenue, and fan loyalty. The data bears this out. Trust is fickle. Once you lose it, getting it back is three times harder than it took to gain it in the first place.

2. **Prevention Is Easier Than Damage Control**

   - Addressing issues early prevents them from spiraling into crises. Never, ever just hope that things just go away on their own. Avoid them if you can, sure. If they do arise, however, make them a priority and get after them.

3. **Opportunities Depend on Credibility**

- A tarnished reputation can dissuade potential collaborators and sponsors. The calculus involved in a brand marketing meeting is simple: will this creator make us look good or will they make us look bad? No nuance in those meetings. Yes or no. You do not get a second chance to make a first impression.

Step 1: Building a Secure and Professional Online Presence

1. **Choose Your Platforms Wisely**

   - Focus on platforms that align with your content goals and target audience. Diversify your presence to reduce dependency on any single platform.

2. **Secure Your Accounts**

   - Use strong, unique passwords for each account.

   - Enable two-factor authentication (2FA) to prevent unauthorized access.

   - Monitor login activity and revoke access to unfamiliar devices.

3. **Use Verified Accounts**

    - Obtain verification badges on platforms that offer them. This adds credibility and reduces the risk of impersonation.

4. **Professional Branding**

    - Use consistent usernames, profile pictures, and branding across platforms to build recognition.

    - Craft a bio that clearly communicates your mission and values. Short and sweet.

Step 2: Protecting Your Content

1. **Copyright and Trademark Protections**

    - Register copyrights for original content like videos, artwork, or blogs.

    - Trademark your brand name, logo, or slogans to prevent misuse.

2. **Watermark Your Work**

    - Adding watermarks to photos and videos deters unauthorized use and ensures proper attribution.

3. **Monitor for Unauthorized Use**

- Use tools like reverse image search (e.g., Google Images or TinEye) and content monitoring services to identify unauthorized use of your work.

4. **Use Platform Tools**

- Platforms like YouTube and Instagram offer built-in copyright management tools to claim and monetize your content when reused.

Step 3: Preventing and Addressing Defamation

1. **What Is Defamation?**

- Defamation involves false statements made publicly that harm your reputation. We will get into that a bit later in Chapter 12. Long story short, defamation can be:

    o **Libel:** Written defamation (e.g., blog posts, tweets).

    o **Slander:** Spoken defamation (e.g., podcasts, live streams).

2. **How to Prevent Defamation**

- Be mindful of how you engage with controversial topics or critics. Avoid aggressive or inflammatory responses that might escalate conflicts. Social media "tirades" and overly political commentary, while satisfying, will hinder the value of your brand. How many sponsorships does Mark Ruffalo have? Not many. How about John Cusack or James Woods? Probably none. I'm sure they're all perfectly pleasant human beings in person, but their online presence could be considered divisive and toxic for mainstream brands. Got strong opinions? Great. Just understand it may limit your earning potential.

3. **Responding to Defamation**

- Gather evidence, including screenshots, timestamps, and links.

- Demand corrections or retractions from the offender or website, as appropriate.

- Consult a lawyer to evaluate your options, including cease-and-desist letters or lawsuits.

Step 4: Managing Cybersecurity Risks

1. **Recognizing Common Threats**

    •   Phishing attacks aimed at stealing your login credentials.

    •   Malware that compromises your devices and data.

    •   Social engineering tactics designed to trick you into revealing sensitive information.

2. **Cybersecurity Best Practices**

    •   Use antivirus software and keep it updated.

    •   Be cautious about clicking on links or downloading files from unknown sources.

    •   Regularly back up your data to a secure location.

3. **What to Do If You're Hacked**

    •   Change your passwords immediately and enable 2FA (if you haven't already).

    •   Notify your audience if your accounts were compromised and warn them about potential scams.

- Contact the platform's support team to recover your account and report the incident.

Step 5: Handling Online Controversies and Backlash

1. **Stay Calm**

   - Avoid reacting impulsively. Take time to assess the situation before responding.

2. **Evaluate the Validity of Criticism**

   - Determine whether the criticism is valid and if an apology or clarification is warranted. Personally, shutting up for a while is more valuable than issuing an apology at any time.

   - You may be directed to issue an apology by a commercial partner or risk losing them – don't just buckle to that pressure. Assess the criticism and take some time to do it. Be honest about it and be objective.

   - Reflexively bending to the whim of the mob does not guarantee anything will change or even that you will get yourself out of trouble. Trolls are real. Outrage is often manufactured for effect. Know those things and, when it comes time to decide, make a decision that you feel (objectively) makes sense for you and your brand. Authenticity is more valuable than fake contrition.

3. **Engage Strategically**

- Issue a thoughtful response, if appropriate, that addresses concerns without escalating the issue.

- Avoid getting into public arguments, especially on social media. Going dark is usually the smarter play.

4. **Leverage Your Support Network**

- If you have a PR team or legal advisor, involve them in crafting a response. Outside advisers will have a different take on the issue, I can guarantee. It's not personal to them, even though you might think it's the end of your universe.

Real-World Examples of Reputation Management

1. **The Redemption Arc**

- A popular influencer faced backlash for offensive tweets from years ago. By issuing a sincere apology and committing to actionable changes, they regained audience trust over time - slowly.

2. **The Impersonation Scandal**

- A creator's identity was impersonated on multiple platforms. By reporting fake accounts and addressing the issue publicly, they mitigated the impact and regained control of their narrative.

3. **The Cyberattack Recovery**

- A YouTube channel was hacked, and all videos were deleted. The creator's swift action to contact platform support and communicate with fans minimized the damage and led to account recovery within days.

Actionable Tips for Long-Term Reputation Management

1. **Monitor Your Online Presence**

- Set up Google Alerts for your name or brand to stay informed about mentions.

2. **Engage Authentically**

- Build genuine connections with your audience to foster goodwill and resilience against negative incidents.

3. **Have a Crisis Plan**

- Prepare a checklist for handling account hacks, negative press, or public controversies.

4. **Invest in Professional Help**

- Consider hiring a PR specialist or reputation management firm if your brand faces ongoing challenges.

NERD ALERT: Let's drill down a bit on "crisis management." If you ask 10 experts about how to handle a reputational crisis, you'll get 11 different opinions. None of those opinions matter more than your own – including this one. Ultimately, the face looking back in the mirror is the one you need to live with.

As someone who is involved in knock-down, drag-out disputes all the time, a fair amount of name-calling and insults are thrown around. I've been called every name in the book, and sometimes names that aren't even in the book. My advice is this: let it go. Take some time to let things cool down.

There will be a lot of talk about "trying to get out in front of things." That never works. I'm also not a fan of apologizing simply because someone demands one. All that does is embolden them to come after you again or to dig in deeper and demand more. They want scalps and compliance. If you're not going give them either one, they'll move on to another sucker. Never apologize in the middle of a crisis, either, because it only fuels the flames. Feel free to speak privately with people you feel need more context or your side of the story, but nothing publicly. Those who need to know, get to know. No one else.

There's a Biblical proverb that says, "this, too, shall pass." Whether or not you are a believer, that was really kick-ass crisis management advice, especially during the thick of a crisis. Time has a wonderful effect on cooling everything off. Let it do its job. After a few days, dust yourself off, move forward, and make decisions that you can live with.

Life is long, clicks disappear, and popularity is fleeting. Find some meaning in what you do online, pass along your wisdom to others, expand your network, and strive to be better every day. If you do those things, you'll find that the disputes are fewer, easier, and far between. You'll find that your success is even sweeter. And, when you close your eyes at night, you'll just sleep. I promise you, that will be more valuable than anything you can do online or any checks you can cash. Sermon over. Good luck.

Protecting Your Legacy

Your online presence is one of your most valuable assets as a creator. Your integrity and professionalism are two more. By proactively managing your reputation, securing your accounts, and preparing for potential challenges, you'll not only protect your brand but also set the stage for long-term success.

In the next chapter, we'll explore issues surrounding collaboration with others, i.e., working and playing well in the sandbox.

CHAPTER 10

Collaborating with Others & Sharing IP Safely

The Power of Collaboration

Collaborating with other creators, brands, and partners can elevate your work (and earnings) to new heights. Whether it's a joint YouTube series, co-creating a product line, or teaming up on a sponsored campaign, partnerships can expand your audience, diversify your content, and generate new revenue streams. However, just like joining a middle school cheerleading squad, collaboration also brings potential risks, especially when IP is involved.

In this chapter, we'll explore how to collaborate effectively while protecting your rights and ensuring your partnerships and "collabs" are mutually beneficial.

Step 1: Define the Scope of the Collaboration

Clear communication at the beginning is essential to avoid misunderstandings. Before jumping into any partnership, establish the following:

1. **Objectives**

   - What are the goals of the collaboration? Are you aiming to reach a new audience, share resources, or co-create content?

   - Just messing around is a completely fine "objective," too. Some of the best things happen in the moment. Not everything has to be so freakin' serious.

2. **Roles and Responsibilities**

   - Who is responsible for what? For example, if you're co-creating a video, clarify who will handle filming, editing, and promotion. Be fair and leverage efficiencies.

3. **Ownership of Work**

   - Determine who will own the IP created during and after the collaboration ends – voluntarily or not. Will it be a joint ownership (with each party retaining a right to use), or will one party retain exclusive rights?

Example: If two influencers create a music video together, decide whether both have equal rights to monetize and distribute it or if one party will take the lead.

Let's Talk About Term Sheets. Ideally, though often overlooked, the easiest way to address (and expose) potential down-the-road problems is to create a simple, often one-page, term sheet.

A good term sheet identifies the WHO-WHAT-WHERE-WHEN-WHY aspects of the project. If you both cannot come together on a basic term sheet, the chances of successfully drafting a detailed contract you both agree on or, worse, performing in a way that both parties expect under that detailed contract are remote (read: not going to happen).

Before starting the project, you absolutely need to agree in principle the following:

- Who's contributing what? Money, time, music licenses, equipment, preexisting IP, the cat, microphones, gas money, skimpy swimsuits, makeup, snacks, etc. Figure out a division of labor for each step of the way. Don't forget promotion, editing, or cleaning up your mother's basement. Disputes happen almost immediately if parties think they are being taken advantage of and the other is not pulling his weight.

- What is the project going to look like at the end? You'd be surprised how often this gets overlooked. Give it a fancy project name if it helps keep people on task. OPERATION: RIZZ SIGMA. You know, something cool like that.

- Where does the "product" go at the end? Your page, their page, both pages, multiple platforms, etc. Is there a schedule?

- When does the project start and, importantly, end? If you want to be partners moving forward on all things, great. State it out loud. If you are coming together for a one-off collaboration, also fine. Make sure everyone is clear right up front.

- Why are we doing it? Likes, clicks, referrals, money, clout, free pizza, whatever. How do the rewards get split and when?

These questions need to be answered BEFORE you even bother burning calories on doing a contract. When you call your lawyer, he will ask all of these questions. Have an answer or not; he's getting paid by the hour while you figure out an answer. Have a term sheet with all of his questions answered before making the call? He'll send you a bottle of wine at Christmas and tell all of his lawyer buddies about the unicorn he just spoke to on the phone.

Step 2: Use Written Agreements

Oral agreements might work for casual or in-the-moment collaborations, but for anything substantial, a written contract is non-negotiable. In line with what you decided in the term sheet, your agreement should (and better) cover:

1. **IP Ownership and Licensing**

   - Specify who owns the content and whether any licenses are granted for use by either party. Make sure that if you contribute any preexisting IP to this project, you have the rights to do that, as well.

   - Include terms for shared or joint ownership if applicable.

2. **Revenue Sharing**

   - Clearly outline how profits (e.g., ad revenue, merchandise sales) will be divided. Also important is to state when those profits will be distributed. Upfront? At the end? 30 days after getting any money? Whatever it is, make sure you cover it off in the contract.

3. **Confidentiality**

- Protect sensitive information shared during the collaboration, like business strategies or unreleased projects. One typical term that is often confidential is the fact that you are doing a collaboration at all – at least until you're ready to tell the world or, stated differently, until you look good in the video and the edit is strong.

4. **Dispute Resolution**

- Include a method for handling arguments and disagreements, such as mediation or arbitration. We will discuss this in more detail in Chapter 12. If you have a trusted third-party who you both feel confident can be fair if a dispute happens, make it mandatory that you refer all issues to them before doing anything else. This can help keep the legal costs way down. But, you both would need to trust Becky and, ideally, Becky isn't a 12-sided die.

Example Clause: "Both parties agree that any revenue generated from the project will be split 50/50 after deducting shared expenses. In the event of a dispute, both parties will attempt in good faith to resolve the dispute through mediation before pursuing legal action."

Step 3: Protect Your Intellectual Property

Collaborations often involve sharing your existing IP, such as:

• Logos, trademarks, or branding.

• Pre-recorded videos, scripts, or music.

• Proprietary tools or processes.

To safeguard your assets:

1. **Grant Limited Licenses**

    • Allow collaborators to use your IP only for the scope of the project. Specify how, where, and for how long it can be used. This can be done in the same agreement – no need for two contracts.

2. **Retain Ownership**

    • Make it clear that sharing IP does not transfer ownership. For example, if you share a logo for a co-branded event, the collaborator can't use it outside that context or in any way that you don't authorize in writing in advance.

3. **Monitor Usage**

- Keep an eye on how your IP is being used during and after the project to ensure compliance.

Real-World Example: A podcaster partners with you, a graphic designer, to create cover art for the podcast. The agreement states that you retain copyright but grants the podcaster an exclusive license to use the artwork to promote his show. If the podcaster turns around and uses the cover art on a bespoke wine venture with his cousin in Santa Cruz, you might need to discuss what happens next. Is it promotion? Arguable. Is it promoting his show or his wine? Debatable. Is he treating the artwork like he owns it? Perhaps.

Be clear how things can be used before they become a problem.

Step 4: Addressing Joint Ownership

Not unlike the example above, joint ownership can be complicated, as it requires ongoing cooperation. If you decide to jointly own IP, consider:

1. **Usage Rights**

   - Can each party use the IP independently, or do they need mutual consent?

2. **Revenue Splits**

   - How will profits from the IP be divided? Will both parties contribute equally to its monetization? It's not always 50/50. Some people may contribute money instead of talent to create the IP. They may be entitled to a larger portion of the pie.

3. **Exit Strategy**

   - What happens if one party wants to end the collaboration? Will they sell their share, or will the IP be dissolved? Will there be a forced buyout option (e.g., for some insane amount of money, the other party agrees to sell their share, no questions asked).

Step 5: Collaborating on Sponsored Content

When working with brands or sponsors, ensure clarity on:

1. **Content Guidelines**

- Understand the brand's expectations for tone, messaging, and deliverables.

2. **Disclosure Requirements**

- Follow platform and FTC rules and legal guidelines for disclosing sponsored content. #ad.

3. **Exclusivity**

- Some sponsors may require you to avoid promoting competitors during the campaign. It is a common provision, for example, for endorsers of one type of beer to not be seen publicly with another brand in their hands (there's a famous story about Jimmy Kimmel refusing a certain kind of beer on the back nine of a golf outing because he endorsed another brand, much to his disappointment).

- Confirm the duration and scope of such clauses. Fair warning, sponsors have no interest in helping you out. They care only about moving units and looking good to their bosses. As such, they will be pigs and try to secure rights they don't deserve and probably are not paying for. Like exclusivity rights in perpetuity (meaning forever). If you see that word, grab your little sister's fattest red crayon and circle it on the contract. Then, send it to your lawyer. Then, see if their competitor is making a better deal.

Example: A beauty influencer collaborates with a cosmetics brand on a product launch. The agreement specifies a 6-month exclusivity period during which the influencer cannot promote competing brands. This is not too onerous, but you need to ask yourself what it means to turn down offers from competing brands for 6 months. What does "promotion" entail? Using a competitor's product? Also, see how they define what a "competitor" is in the contract. One of the classic ways brands like to act like pigs and overreach, FYI.

Step 6: Managing Collaborative Content on Platforms

Platforms like YouTube and TikTok have specific rules for co-creating content and sponsorships. Keep these in mind:

1. **Revenue Sharing Features**

   - Platforms like YouTube offer built-in revenue-sharing tools for collaborative videos. Use these to simplify profit distribution.

2. **Attribution**

- Ensure all contributors are credited appropriately in descriptions or tags. TikTok even mandates that you use their proprietary tool for labeling sponsored posts.

3. **Content Ownership**

- Verify who owns the rights to uploaded content. Some platforms might claim partial ownership or licensing rights.

Step 7: Handling Disputes and Breakdowns

Despite the best planning, disputes can and will arise. Here's how to manage them:

1. **Communicate Openly**

- Address issues as they arise to prevent misunderstandings from escalating.

2. **Refer to the Agreement**

- Use the contract as the foundation for resolving disagreements. This is called in the legal world, quoting "chapter and verse." If the contract says no one gets paid until the video goes live, then point it out. They signed it. They agreed to it. They don't have to like it, but they do have to honor it.

3. **Seek Mediation**

- If direct communication fails, involve a neutral third party to mediate the dispute (e.g., Becky, after she gets off of work at Sephora – don't call her at work; that's not okay).

Example: Two vloggers co-create a travel series but disagree on editing styles. Their agreement includes a clause for final decisions to be made by an editor they both trust. Be careful with clauses like this, however. That third editor may not be willing to work for free. If you cannot agree, then you'll both have to pitch in to pay for the neutral party to resolve your issue. These kinds of clauses can drain the profits out of a collaboration quickly.

NERD ALERT: *Let's take a quick look into 2 case studies, both of them involving popular podcasts affiliated with Barstool Sports. Just like Dave Portnoy says in front of pizza joints, "One bite, you know the rules."*

*Case #1 – **Call Her Daddy**. Before it became the preeminent ladies' potty-mouthed, slut-glorifying juggernaut, the Call Her Daddy podcast IP was owned by Barstool Sports,* including the name and content, under the terms of its contract with hosts Alex Cooper and Sofia Franklyn, the podcast's original founders. As the popularity surged (and Cooper got a powerful new boyfriend), the hosts demanded more favorable terms around creative control, ad revenue sharing, and IP rights. After private negotiations led to a public fallout, Franklyn was out. Cooper negotiated a new deal with Barstool's owner, Dave Portnoy, to continue the show for a period as a solo host for a reduced advertising share. But, smartly, Cooper secured the ownership rights to the IP, including the name. This was a win-win for both parties. Barstool got to keep wringing out some short-term revenue from an unhappy brand and host. Once she fulfilled her term, Cooper took the IP and the show to Spotify for an insane $60 million deal.

Case #2 – **Bussin' with the Boys.** Another wildly popular podcast hosted by former NFL players Taylor Lewan and Will Compton, took a slightly different (smarter) path. When they contracted with Barstool Sports, they retained their IP rights, including the name and content, from the very beginning. By doing so, they ensured they could operate independently if things went sideways. The agreement allowed them to grow their brand by aligning with Barstool's platform for promotion and distribution. Barstool got a blue-chip podcast against which to sell a ton of advertising. As these things go, it appears that, as of January 2025, the parties will be going their separate ways. Lewan and Compton have indicated that they will remain independent and not tied to a particular platform, though that is doubtful given the news *reports I've seen. If someone is being asked to match a deal (or come close), there's another deal out there. Time will tell, but the lesson is a good one: keep your IP if you can. You never know when you're going to wake up one day and need it. And, having to buy it back from someone will make your stomach churn for a very long time.*

Don't cry for any of these folks. They're all doing just fine.

Actionable Tips for Successful Collaborations

- **Start Small.** Begin with smaller projects to build trust and gauge compatibility before committing to larger ventures.

- **Document Everything.** Even informal collaborations benefit from e-mail summaries (essentially a term sheet) and written agreements.

- **Be Transparent.** Share expectations and concerns upfront to avoid surprises.

- **Respect Boundaries.** Understand and honor your collaborator's creative vision and IP.

The single best way to get more collaborations and to have successful ones is to be a good partner. Be open about your expectations, give it your best effort, and don't be petty. A lot of promising partnerships have been thrown away out of just plain pettiness. Don't get caught up in using the word "I" when you should be saying "we." Classic red flag. Don't think that your contributions are somehow more valuable simply because they came from your precious imagination. Remember why you went into partnership in the first place. So, you both could get something out of it and you both would succeed. Remember that?

Don't be the douchebag who decides to ditch his partner, take all the credit, and accept all of the accolades just when the venture is starting to succeed. Usually, it's for money. Often, it's about ego. Always, it's petty and short-sighted. Content creators have no monopoly on douchebag partners, believe me. Every industry has them. Just don't be one. That's all. You'll live a better life and have a lot more success in the end.

When your name comes up on somebody's phone, you want that person to pick it up with a measure of excitement. You don't want them to press the "Voicemail" button because the thought of talking to you is tedious, counter-productive, and painful.

Collaborate Without Compromise

Collaborations can be transformative for creators, offering fresh ideas, expanded reach, and increased revenue. They can also be insufferable and littered with narcissistic turds who want to rip you off and toss you out in the street the first chance they get. Often, they are somewhere in between those two extremes.

By setting clear expectations, protecting your IP, and fostering open communication, you can navigate partnerships confidently and avoid common nightmares. Be generous with your time and your praise, keep in mind why you thought a collaboration was a good idea to begin with, and generally act in a way that makes other people want to do more great stuff with you in the future. Get up, show up, stay up, clean up. That's a recipe for success, right there. [For those of you unfamiliar with that certain latter day Will Rogers, this where Larry the Cable Guy would say, "Git-R-Done."]

In the next chapter, we'll some basics around contracts and agreements. This is the part you've been waiting for, really. I know. Me, too.

CHAPTER 11

Understanding Contracts & Agreements

Why Contracts Matter

Contracts are the backbone of professional relationships. They establish expectations, protect your rights, and help prevent misunderstandings. Whether you're hiring a designer, collaborating with another creator, or signing a brand deal, understanding contracts is essential for safeguarding your interests. Despite the flashy intro about the lofty importance of contracts, lawyers who specialize in contract law are among the most boring people you will ever meet. No lie. Present company included.

Just off the top of my head, content creators (at some point or another in their journey) may encounter: confidentiality agreements, purchase and sale agreements, partnership agreements, website terms of use, privacy policies, end user license agreements, software agreements, platform TOS, e-commerce agreements, banking agreements, merchant processing agreements, distribution and licensing agreements, shopping agreements, option agreements, management contracts, independent contractor agreements, endorsement agreements, vendor agreements, rental agreements, and sponsorship agreements. We are definitely not going to cover all of those, don't worry.

In this chapter, we'll break down the key components of contracts generally, discuss a few in more detail, provide real-world examples, and offer some tips to ensure you're signing agreements that work in your favor.

What Is a Contract?

At its core, a contract is a legally binding agreement between two or more parties. It outlines, in a nutshell:

1. **Who:** The parties involved.

2. **What:** The obligations, services, or products being exchanged.

3. **When:** Timelines and deadlines.

4. **How:** Payment terms, deliverables, and expectations.

5. **What If**: Remedies for breaches or failure to meet terms.

Contracts can be written, verbal, or implied. However, written contracts are the most reliable and enforceable. If you are hanging your hat on the enforceability of an oral agreement, you're already behind. And, there may not be enough time left in the game to catch up. Handshake deals and gentlemen's agreements are myths made up by your worthless uncle who wishes it was 1962 again. Be a proud member of this (smarter) generation and get it all in writing.

If you run into a potential partner who "doesn't like to do things too formally" or "doesn't need a piece of paper because his word is his bond" or "thinks it's too big a hassle to get lawyers involved," run – do not walk – away. They are either: 1) illiterate; 2) your ex-boyfriend; 3) not serious; or 4) going to rip you off. No matter which bucket he falls into, drop him like a hot rock and find someone who wants to act like an adult. This also works in the dating world, by the way. Relationships are just contracts, after all (but the remedies for breach can be super-messy). Which is why, incidentally, I do not practice matrimonial law.

NERD ALERT: If you've made it this far in the book, I presume you have found it helpful and not overly technical in tone or language. In case you think it's still too "technical," let me share with you what most books about the law sound like. Just wholesome old-timey law-talkin'. Here is one eminent scholar's definition of the word "contract":

> *"The term contract has been used indifferently to refer to three different things: (1) the series of operative acts by the parties resulting in new legal relations; (2) the physical document executed by the parties as the lasting evidence of their having performed the necessary operative acts and also as an operative fact in itself; (3) the legal relations resulting from the operative acts, consisting of a right or rights in personam and their corresponding duties, accompanied by certain powers, privileges, and immunities. The sum of these legal relations is often called 'obligation.' The present editor prefers to define contract in sense (3). ..."*

> *Source: Willam K. Anson, Principles of the Law of Contract, p. 13, note 2 (Aruthr L. Corbin, editor, 3d Am. Ed. 1919).*

Pretty dynamic stuff, right? I told you contract lawyers were dull.

Key Components of a Contract

Every solid contract includes these essential elements:

1. **Offer and Acceptance**

   - One party proposes terms (the offer), and the other agrees (acceptance). The terms must be accepted exactly as offered, with no variation whatsoever. If you don't accept them as offered, the offer is dead, and you have just made a counter-offer. Which needs to be accepted exactly as offered...and so on.

2. **Consideration**

   - Each party must exchange something of value, such as money, services, or goods. A promise is also valid consideration. For example: If you paint a picture of my pet goat, Stumpy, I promise to pay you $500 – that promise is perfectly valid consideration.

3. **Legal Purpose**

   - The agreement must be for a lawful activity.

4. **Capacity**

- All parties must have the legal ability to enter into a contract (e.g., being of sound mind and of legal age).

5. **Mutual Consent**

- Both parties must agree to the terms without coercion, duress, or fraud.

Types of Contracts Content Creators Encounter

1. **Influencer Agreements**

- Agreements with brands for sponsored posts, product reviews, or partnerships.

- *Example Clause*: "Creator will post one (1) Instagram story and one (1) feed post featuring Brand's product within seven (7) days of receipt."

2. **Collaboration Agreements**

- Contracts between creators working together on content or projects.

- *Example*: Two YouTubers co-creating a video series and splitting ad revenue.

3. **Service Contracts**

- Agreements with freelancers like graphic designers, editors, or managers.

- *Key Point*: As with any contract, specify deliverables, payment terms, and deadlines. You can create generalized templates that leave open blanks to fill in for some of these simpler vendor contracts. They come in handy. But, only if you actually use them!

4. **Platform Agreements**

- Terms of service (TOS) for platforms like YouTube, TikTok, or Patreon. You may not have read them, but you should. They are contracts, and you will be bound to their terms.

- *Example*: Monetization rules, content restrictions, copyright enforcement, and revenue-sharing percentages.

5. **Licensing Agreements**

- Contracts granting permission to use someone else's IP or your granting permission for others to use yours.

- *Example*: Using copyrighted music in a video.

The Loose Language of Licensing. Of the group, licensing agreements are probably the trickiest ones because people write licenses with all sorts of insider/lawyer language and vernacular. At their bottom, though, think of them like hall passes. If you want to go to the bathroom during class, you need a pass from the owner (i.e., the teacher). Think of the platforms as the hall monitors. They may not catch you, but, if they do, you better have the hall pass.

I will try to demystify some of that coded language below, so you understand the basic concepts, at least. As I mentioned above, every time you upload content to YouTube or Facebook or whatever platform you prefer, you are actually granting them a license (i.e., permission) to use your content.

Here's the language from the YouTube TOS:

"By providing Content to the Service, you grant to YouTube a worldwide, non-exclusive, royalty-free, sublicensable and transferable license to use that Content (including to reproduce, distribute, prepare derivative works, display and perform it) in connection with the Service and YouTube's (and its successors' and Affiliates') business, including for the purpose of promoting and redistributing part or all of the Service." Source: https://www.youtube.com/t/terms

Let's break that word salad down a bit.

First, notice the capitalized words ("Content," "Service," etc.). They're not just trying to be fancy about it. As a general rule, capitalized terms are *defined* terms, laid out somewhere earlier in the contract. The term "Service" in this agreement, for example, was earlier defined as "the YouTube platform and the products, services and features we make available to you as part of the platform." Every time you see the capitalized term in a contract, it means the same thing as the first time they used it (and defined it).

Second, note that this provision has you granting YouTube a license (i.e., permission). Not the other way around. What kind of license, you ask? Good question. A "worldwide, non-exclusive, royalty-free, sublicensable and transferable" license. That clears everything up, right? Not hardly. Let's go deeper:

- **"worldwide"** – that means pretty much what it says – they can use your content anywhere;

- **"non-exclusive"** – that means you can also license your content to other platforms;

- **"royalty-free"** – that means they're not going to pay you anything for the permission to use your content;

- **"sublicensable"** – that means they can take your permission and create another license with it to a third party (there are some limits on how they can do that);

- **"transferable"** – that means they can take your permission and give it/sell it/loan it to another entity to use your content (again, some limits on this).

Third, this clause also tells you a bit about how they can use the content you have just given them permission to use for free, everywhere. The term "use" is not capitalized, but the context of the sentence gives you its meaning: reproduce, distribute, prepare derivative works, display, and perform it. Basically, anything you can do with it, they can do with it.

Finally, the only real limitation is that the license they received from you can only be used "in connection with the Service." Recall the definition of Service (above). While not in that definition technically, this clause adds a few wrinkles about what YouTube believes the definition of "Service" includes – "promoting and redistributing part or all of the Service." So, instead of just limiting their use, they take a few minutes to expand their own definition, as well. Just like when your little brother takes the last Oreo while you are checking your texts.

And, it's not just YouTube. You must feeling super-generous now because the next paragraph in the agreement requires you to give a license to *every other user* on the platform, too. Yeah! So long as those users use your content "as enabled by a feature of the Service" – whatever the hell that means. Honestly, that could mean almost anything.

What are you going to do about it? a) Not use YouTube; or b) suck it up and live with it. Those are your only options. Not exactly the "mutual back-and-forth" you had in your mind about contracts, right? More like "take it or leave it." But, let's not just bash YouTube. They all do it. Plenty of pigs at this trough. Looking at you, Zuckerberg.

Red Flags to Watch For

When reviewing contracts, be cautious of the following:

1. **Ambiguous Language**

   - Phrases like "subject to approval" or "in the sole discretion of" can give the other party too much control. Parties are always trying to grab more advantage by using language that seems innocuous at first but which will be devastating or, at the very least, surprising to you down the line.

- Don't let them give you some lame excuse as to why "we need that language" or why "corporate requires this." If they need it, they should pay for it. When they have to pay for it, they often discovery they don't really "need" it.

2. Unreasonable Exclusivity

- Clauses that prevent you from working with competitors for an excessive time or scope. Be extremely careful with these clauses. They want you to sit on the sidelines and not work with their competitors. Which is fine. But, that may stop you from doing anything at all for months or years – especially if you are industry-specific (e.g., makeup, fashion, food, etc.). As above, if they want something, they will pay for it. You can agree or not.

3. No Termination Clause

- Contracts should allow for termination under specific conditions. Ideally, all contracts should have a natural termination and, if everyone is working well together, mechanisms to extend. "Perpetual" clauses in contracts are never good.

4. **Unfair Indemnification**

- Clauses requiring you to cover all legal costs in case of a dispute, even if it's not your fault. Absolutely not. Never cover their legal costs. Everybody should bear their own attorneys' fees. That's how it works in the U.S. Anyone who says differently is trying to gain advantage and hoping that you're not paying attention.

5. **Lack of Payment Details:**

- Ensure payment amounts, methods, and timelines are clearly defined. This is super-important. Go right to the payment provisions before looking at anything else. Bear in mind that some large corporations have some processes and procedures they need to follow to get people paid. 30 days to get your money is pretty standard. 10 is bordering on unreasonable. 3 days is impossible, usually. If you're going to have issues with payment timing, try to get some money up front and the rest later.

Breaking Down Complex Terms

Contracts always contain lawyer-speak and legal jargon. Here's a quick glossary:

1. **Force Majeure**

   - From the French, meaning "superior force."

   - Frees parties from obligations due to events beyond their control (e.g., natural disasters, wildfires, lightning strikes, other acts of God, terrorist attacks, manufactured worldwide pandemics, etc.).

2. **Indemnification**

   - One party agrees to cover losses or damages incurred by the other.

3. **Non-Compete Clause**

   - Restricts you from working with competitors for a specified time and, often, in a specified region. These are often egregious and, so you know, difficult to enforce.

4. **Confidentiality Clause**

- Prohibits sharing sensitive information. Bear in mind that the contract itself is very often considered "confidential" and the fact that it even exists should not be shared around town.

5. **Liquidated Damages**

- Pre-agreed compensation for breaches, like missing a deadline. These are almost always unenforceable and people use these clauses as penalties to keep you from leaving. If you see these kinds of clauses, be very wary of your new contract partner. They've been burned in the past, and definitely have trust and/or daddy issues. It's the legal equivalent of a tramp stamp.

Real-World Examples of Contract Disputes

1. **The Unpaid Freelancer**

- A creator hired a video editor without a written contract. When the editor delivered subpar work, the creator refused to pay, and the editor took legal action. This will be a mess because you didn't identify the rules of the game before rolling the dice. One tip – offer the editor something but only on the condition he doesn't sue or drops his suit (this is called settlement in "full satisfaction"). If he takes it, dispute is over. If not, didn't cost you anything to try. But, get it in writing, either way! E-mail is fine.

- *Lesson*: Always define quality standards and payment terms in writing – before starting any work.

2. **The Influencer Backlash**

- An influencer signed a brand deal but failed to disclose it was sponsored content. The platform suspended their account, citing breach of terms. This is just dumb, and you should know better. Guess what? You might get sued by the brand for failing to honor your end of the bargain. That's a double-whammy.

- *Lesson*: Read and follow platform agreements.

3. **The IP Ownership Tug-of-War**

- Two creators co-wrote a song but didn't establish ownership. When it went viral, they disagreed on royalties. This one will end up involving lawyers if you two don't solve it quickly and fairly.

- *Lesson*: Address IP ownership up front – right after payment terms.

Negotiating Contracts Like a Pro

1. **Do Your Research**

   - Understand industry standards for rates, rights, and obligations.

2. **Ask for Changes**

   - Contracts are negotiable. Don't hesitate to request edits. If they push back, ask them to explain why they "need" the language as-is. Usually, they refer that to "legal" to explain. Either way, this is your deal with them; you're not at the mercy of what some prior idiot agreed to. If they want you bad enough, the legal department will get off their backsides and tweak the language to something on which you both can agree.

3. **Get It in Writing**

   - Verbal agreements are hard to enforce. Confirm all changes in writing.

4. **Consult a Lawyer**

- For complex agreements, professional legal advice is invaluable. Until you have some experience under your belt, get some legal advice on the simple agreements, too.

- *Remember*: Lawyers have seen everything. Nothing surprises us, and we don't take contract language personally. We know the magic words that other lawyers insert to gain advantage over you. Those are the kinds of people you want on your side.

Key Clauses to Include in Every Contract

1. **Scope of Work**

   - Clearly define deliverables, deadlines, and responsibilities. Who is doing what? By when? What happens if they mess it up? Need to consider that very likely scenario, I'm afraid.

2. **Payment Terms**

   - Specify amounts, methods, and timelines.

3. **Termination Clause**

- Outline conditions for ending the agreement. A definite "term" (i.e., how long the contract lasts) should be included. Try to avoid leaving things open-ended if you can. "Perpetual" or "in perpetuity" are each a no-go. Always.

4. **Dispute Resolution**

- Include steps for handling disagreements, such as mediation or arbitration.

- *Pro tip*: have all disputes escalated to a senior manager before any formal dispute resolution process (litigation, mediation, or arbitration) is started. I will discuss this in more detail later. Saves a lot of money and time.

5. **Confidentiality and IP Ownership:**

- Protect sensitive information and clarify who owns the work. Even if the brand, for example, wants to own the IP at the end, demand a no-royalty (i.e., free) license from them so that you can use it on your page, in your portfolio, etc. Recall the YouTube TOS example from before. Turn it back on them.

Actionable Steps for Creators

- **Always Review Before Signing**

  o Take your time to read and understand every clause. This is easier said than done, obviously. Get help if you need it. No shame in trying to learn about complicated things. I wouldn't have the foggiest idea how to apply a filter to make my face look less fat. But, I would ask if I needed something like that. ☺

- **Use Templates Wisely**

  o Customize contracts to fit each situation. Work with a lawyer who understands that you need some basic template contracts without breaking the bank. Confidentiality agreements, basic independent contractor (freelancer) service agreements, IP license agreements/clauses, release agreements, etc. There's one lawyer I know who makes these things available at a reasonable price: www.titanlawny.com #attorneyadvertising.

- **Maintain Records**

  o Keep copies of all agreements and related communications. All disputes come down to documents and whomever keeps better files usually wins.

- **Leverage Tools**

  o Use e-signature platforms like DocuSign for convenience and legality. These are great, and they save time. You almost never need an original signature any more. Also, be sure to download the platform's proof of signature and verification as to when and how the document was accessed, reviewed, and executed.

Contracts Are Your Second Best Friend

Your contract attorney is your best friend, actually. But, when he's out skiing in Aspen, your ability to handle basic contracts will serve you well. Understanding contracts might not be glamorous, but it's crucial for long-term success as a content creator. A well-drafted agreement can prevent disputes, protect your rights, and set the stage for successful partnerships. Don't be intimidated and, most importantly, don't ever hesitate to demand an explanation of a term, ask for more clarity in the language, or try to make things "mutual" (e.g., whatever they demand, just say both parties should have that benefit, especially on things like audit rights, termination, and IP licenses).

In the next chapter, we'll explore how to handle things when the crap really hits the fan.

CHAPTER 12

Navigating Nasty Disputes & Legal Challenges

The Ugly Reality of Disputes in Content Creation

No matter how careful you are, legal disputes and challenges can (and often do) arise. From copyright claims to breaches of contract, conflicts are an unfortunate reality for many creators. The good news? You're reading this book, so you're clearly smart. More good news? Knowing how to handle these situations can make the difference between a minor inconvenience and a career-derailing or company-killing event.

Let's paint a picture. Everything with your former collaborator was rosy, fulfilling, and productive. You worked on a few projects together and momentum was building. Then, he started using the terms "I" and "me" and "my" a lot more than was called for. Then, you find out he's been having conversations with other collaborators. Then, the money you were counting on to pay your rent doesn't hit your bank account. Then, you start getting irate calls from brands, vendors, and customers. Then, you find out the money that was earmarked for music licensing was never spent on that. Then, he has a new car. Then, he just stops showing up. Then, legal papers and process servers start showing up at your doorstep. You're pissed off, he's gone, the money's gone, and you want to sue anyone and everyone. Today!

I hope that doesn't sound familiar, but I can assure you it is a lot more common than you might think. If it does ring true, you're smack dab in the middle of litigation now. Welcome to the club. Get ready for a bumpy ride. And, that's the good news. Now the bad news...

Litigation is ridiculously expensive, takes a long time, and can occupy more of your time than you can even fathom. It is used offensively, strategically, and sometimes even frivolously. Some claims are justified; others are just BS. No matter what, if you become part of active litigation (for or against), clear your social calendar for the foreseeable future. No more lattes or girls' night with the "squad" for a while. You might even want to take a spin around Only Fans to see if you need to pick up a side-gig selling feet pics to pay for it all.

Remember the rule of thumb: take what you think it should cost, then double it, then double it again. Depending on the claims of the case, that rule of thumb calculation might be a wild understatement.

I have been a litigator for two decades, handling hundreds of cases and suits in every kind of court imaginable. I will tell you the exact same thing I tell my clients when they come in, hopping mad, ready to "sue somebody!" It's a few things, actually:

1) Sure, we can certainly sue those guys. Right after your retainer check for $25,000 clears my bank account.

2) That's just a down-payment, by the way – it's going to be way more expensive than that before this is over (which could be years).

3) I charge $600 per hour for litigation, so you do the math on how long that retainer us going to last (hint: it's about a week, and I'm fairly cheap in NYC). If you stop paying my bills, I withdraw from the case, and you can start from square one with another lawyer.

4) How will your life or business change if you win? Or lose? Is it worth the fight?

Invariably, I get this line from the client in response: "Look, there's a principle at stake here. They can't just be allowed to get away with it while we do nothing. I shouldn't have to go broke to get what I negotiated."

And, sure, that's a valid feeling. I'm not going to argue with their feelings. It's not productive.

My reply every time is the same: "Principles, like Lamborghinis, are expensive hobbies. They are never as reliable as you expect. They involve maintenance that will dwarf the initial cost. And, the best place to enjoy them, really, is in your mind."

Some might argue that a litigator who works this hard to convince clients not to litigate is cutting off his nose to spite his face. Those people, to be frank, are probably idiots. They have never litigated a case for years, have never broken the news to a distraught client that the resolution they were praying for is not going to happen, have never spent months working on a case only to have a client check out mentally or stop paying in the middle, and are simply not setting appropriate expectations with their clients. Litigation is an all-consuming thing, mixed with equal parts stress, hard work, and divine intervention. Most people are simply not cut out for it – either as lawyers or as clients.

Like it or not, there is no way to reliably predict how any particular case will end. There are an infinite number of ways a case can go sideways or, worse, down the toilet drain. And, it will always be a lot more invasive and disruptive to your business and your life than you can plan for. I promise you that each of those statements is completely and utterly true.

Which is why lawyers get paid up front, charge a lot, and have vacation houses in Lake Tahoe. They know how the hot dogs are made. Honest attorneys will tell you up front how it's probably going to end and the pain that will be endured during the process. Either way, they're going to cash the checks because they're the only ones guaranteed to make any money in the process.

As with anything else, litigation is a game. Play the game or don't. If you do, play it to win. There are no participation trophies and no juice and cookies at the end. Use me or use another lawyer. Doesn't matter. What you want is someone who understands you, appreciates your hard work, knows what he or she is doing, has a plan, and is worth hanging around for a couple of years.

So long as I'm already just handing out free advice, let's explore some common types of disputes, strategies to handle them, and actionable steps to protect your work and reputation.

Common Types of Disputes Content Creators Face

1. **Copyright Infringement**

   - We discussed this in earlier chapters a bit. This is where you land when the nice note, the nasty letter, and the threat of "legal action" don't work. Or, on the other hand, you called the other side's bluff when they sent those things to you.

- *Elements*: To demonstrate copyright infringement, the party who sues (the Plaintiff) must prove (with actual evidence) the following in order to succeed:

1. **Ownership** of a valid copyright (recall that you must register your copyright with USPTO before you can sue on it).

2. **Copying** of original elements of the copyrighted work. This usually requires you to show the other guy (the Defendant) had access to the work and there are substantial similarities to the protected work.

3. **Unauthorized use** or violation of the owner's exclusive rights. These are defined in federal law – 17 U.S.C. §106 – and include reproduction rights (not reproductive rights), derivative work rights, distribution rights, performance rights, among others.

4. **Damages**. How much money did the Plaintiff lose? Sometimes, you can also

get injunctive relief to stop further infringement (discussed above).

- *Example*: Someone uses your video, artwork, or music without permission, or you're accused of using someone else's work improperly.

2. Breach of Contract

- Everyone signed on the dotted line, gave each other hugs, and started counting their money. Now, weeks or months later, someone is not holding up their end of the deal.

- *Elements*: To establish a breach of contract, the Plaintiff must prove the following in order to succeed:

    1. **Existence** of a valid contract. Look to see that there was an offer, an acceptance, and consideration. No contract? No breach.

    2. **Performance** by the Plaintiff. If you are suing on a contract, you have to prove you performed your part or are ready to perform. If you breached the contract, too? Tough sledding.

3. **Breach** by the Defendant. The bad guy didn't perform as required by the agreement.

4. **Damages**. How much money did you expect to make on the deal? You can also get some other types of relief (specific performance, injunctive relief, etc.) but money is by far the most common remedy.

- *Example*: A brand refuses to pay for a sponsored post, or you fail to deliver agreed content on time.

3. **Defamation Claims:**

- These sort of claims arise when people start calling each other names - especially on the internet where it can be seen by everyone in the world.

- *Elements*: To succeed on a defamation claim, the Plaintiff must prove the following in order to succeed (bear in mind you must meet a higher burden to avoid dismissal in claims like this – you must plead the defamation "with particularity," i.e., be very detailed about the exchange – names, dates, times, etc.):

1. A **false statement of fact**. The key is the assertion of fact must be not true. Truth is a defense in a defamation case. If he called you fat and you're morbidly obese, it's going to be difficult to win (or jog).

2. **Publication**. If the publication was oral, it's called slander. If it was in writing, it's called libel. If it wasn't to a third party, this element is not satisfied. In the example above, calling you fat with no one else around is not going to cut it.

3. **Fault**. The Defendant acted either negligently or maliciously – i.e., they knew or could have known the fact was false and made it anyway.

4. **Harm to Plaintiff's reputation**. The statement exposed the Plaintiff to ridicule, hatred, contempt, or financial loss. Certain types of statements (e.g., accusing someone of sexual misconduct) don't require proof of harm; it is presumed (it's called defamation *per se*).

5. **Damages**. Again, usually money damages, but even things like emotional distress will be compensated with money. In defamation *per se* cases,

damages are presumed (i.e., do not have to be proven).

- *Example*: A negative review or commentary you create leads to allegations of libel or slander. Review the elements (above). Truth a defense? Opinion? No malice? Things like that.

4. **Trademark Disputes:**

- These sorts of disputes arise when dueling trademarks or actions cause confusion in the marketplace about where a product or service is originating, e.g., knock-off Louis Vuitton, Specialman (Superman ripoff), Panburger Partner (Hamburger Helper clone), etc.

- *Elements*: To demonstrate trademark infringement, the Plaintiff)must prove the following in order to succeed:

  1. **Ownership** of a valid trademark. Those rights come from registration with USPTO – remember the ® - or via common law (you have some protection just by having used a brand or mark in commerce).

2. **Priority of use**. The Plaintiff has to prove they used it in commerce before the Defendant did.

3. **Unauthorized use** by the Defendant. In commerce.

4. **Likelihood of confusion** among customers as to the source, ownership, or approval of the goods with the mark.

5. **Harm**. That is, money damages.

- One note about trademark cases: It is possible that, in the course of trying to enforce your trademark, the opposing side will look to invalidate your trademark. No valid trademark, no more case. They will attack the underlying registration and, if the court agrees, you could lose what you spent so much money to secure. It happens. These cases can backfire, big-time.

- *Example*: Two creators use similar names or logos, leading to confusion in the marketplace or legal action.

5. **Platform Policy Violations**

- When your battle with the platform has gotten wildly out of control, you will inevitably need to wind up in court if you want back on the platform. These are mostly breach of contract actions, with all of the chips on the platform's side of the table. There might be a dusting of constitutional claims in there, too (e.g., freedom of speech, association, etc.).

- *Example*: Content removal, demonetization, or account bans for alleged violations of community guidelines.

6. Partnership Disputes

- When one partner's ego gets too big for his hat, and he starts thinking you're disposable, well, sometimes you need to haul his big hat into court.

- Again, these are breach of contract actions (the contract is the partnership or collaboration agreement).

- *Example*: Disagreements with collaborators over revenue sharing or creative control.

Steps to Resolve Disputes

1. Stay Calm and Professional

- Emotional reactions can escalate conflicts. Respond thoughtfully and professionally. Don't admit anything in writing, and don't try to be clever. That always blows up later.

2. **Gather Documentation**

- Collect contracts, e-mails, screenshots, and other evidence related to the dispute. Clear documentation is crucial.

- Don't tape your conversations unless you know the rules in your state about that sort of thing. It's usually inadmissible and always less helpful than you think it will be. Plus, if you don't know the rules, you might hurt your case more than you help it.

3. **Understand the Issue**

- Research the specific laws, policies, or agreements involved. For instance, you may have to finally read the platform's TOS if you are terminated or removed. Try to be objective about it. Is it really a travesty of justice, or did you know you were trying to get away with something and got caught?

4. **Attempt Direct Resolution**

- Reach out to the other party to discuss the issue and seek a mutually beneficial solution. Calmly. Professionally.

- *Example*: If a brand refuses payment, refer them to the relevant contract clause, ask to speak with their legal department, and document your demands in writing.

5. **Engage Mediation or Arbitration**

- For disputes that can't be resolved informally, consider mediation (a neutral party helps negotiate) or arbitration (a neutral party decides the outcome).

6. **Seek Legal Assistance**

- For complex or high-stakes disputes, consult a lawyer with expertise in litigating contract and IP disputes.

Navigating Specific Legal Challenges

1. **Copyright Takedown Notices**

- Platforms like YouTube and Instagram have processes for reporting and disputing copyright infringement.

- Steps to Take:

  ○ File a counter-notification if you believe the takedown was incorrect.

  ○ Provide proof of your ownership, such as registration certificates or timestamps.

  ○ Work directly with the claimant to resolve the issue when possible.

2. **Defamation Accusations**

- To avoid defamation claims, ensure your commentary is factual and avoid making unsupported accusations. Podcasters, take note. When in doubt, try to be funny. Satire and parody go a long way toward avoiding defamation claims.

- If accused:

  ○ Request specifics on the alleged defamatory statements.

  ○ Correct or retract statements if they are genuinely inaccurate.

3. **Contract Breaches**

- Follow the contract's dispute resolution clause, which often outlines specific steps like mediation or arbitration. You must do that first before running to court.

o **Mediation**. Mediation is a voluntary and confidential process in which a neutral third party (the mediator) attempts to bring the parties toward a resolution they can both live with. Mediators assist the parties in exploring options and negotiating an agreement. Mediation is often faster, less formal, and more cost-effective than litigation.

o **Arbitration**. Arbitration is a private, formal process in which a neutral third party (the arbitrator) hears evidence and arguments from each side and renders a decision. It is often used as an alternative to litigation, offering a faster, more cost-effective, and confidential way to resolve disputes. Unlike mediation, an arbitrator has the authority to decide the outcome, similar to a judge in court.

4. **Trademark Conflicts**

• Conduct a thorough trademark search before launching a brand or product.

- If someone claims your logo or name infringes on their trademark:

  o Review their registration details. How close is it? Same industry? Will it cause confusion for customers?

  o Negotiate changes or licensing terms, if necessary.

5. **Platform Bans or Demonetization**

- Appeal the decision through the platform's official channels. The TOS may require arbitration of disputes. They may also require you to sue in a specific location. YouTube's terms of service (TOS), for example, requires you to litigate in state or federal court in Santa Clara County, California. Which means you will need to get a lawyer who doesn't do any work for Google in Santa Clara. Good luck. Courts will enforce these provisions. The U.S. Supreme Court has made it clear those kinds of provisions, while ridiculous, will be enforced according to the words in the contract.

- Provide evidence, such as how your content aligns with the platform's guidelines.

Real-World Examples

1. **The Copyright Claim Backlash**

   - A popular YouTuber faced multiple false copyright claims, resulting in demonetized videos and loss of income. After months of filing counter-notifications and rallying community support, the creator's content was reinstated (but not the money lost).

2. **Defamation Lawsuit Threat**

   - A beauty influencer posted a negative review of a product, leading to a legal threat from the brand. In response, the creator's lawyer demonstrated the review was based on personal opinion, not false claims, resolving the matter.

3. **Partnership Fallout**

   - Two creators launched a joint podcast but didn't outline revenue splits. When the podcast became profitable, disagreements arose, straining their relationship.

   - *Lesson*: Always define roles, revenue sharing, and exit plans upfront.

Preventing Future Disputes

1. **Clear Contracts**

   - Use written agreements for all collaborations, sponsorships, and hires. Define roles, expectations, and consequences for breaches.

2. **Understand IP Rights**

   - Register copyrights, trademarks, and other IP to strengthen your legal standing.

3. **Maintain Professional Communication**

   - Keep detailed records of e-mails, messages, and calls related to your work.

4. **Follow Platform Guidelines**

   - Regularly review terms of service (TOS) to ensure compliance. Don't wait until it's too late to review the TOS. I realize I keep harping on this, but this is the world you live in, so you have to know the lay of the land.

5. **Monitor Your Work**

- Use tools like reverse image search to identify unauthorized uses of your content. Conduct regular trademark searches at USPTO. Keep a close eye on competitors.

Cheaper (or Free) Dispute Resolution Resources

- **Legal Aid Clinics**

  o Many cities offer free or low-cost legal advice for creators and small businesses. Often, these are need-based and designed for low-income folks.

- **Online Tools**

  o Platforms like YouTube's Copyright Match Tool or trademark databases (USPTO) can help you monitor and protect your work. For trademark and copyright issues, the USPTO has a ton of free resources you can check out.

- **Professional Associations:**

  o Organizations like the Copyright Alliance offer resources and support for creators.

o   Local bar associations (i.e., lawyer groups) and law schools offer free clinics or referrals to low-cost providers.

Turning Challenges Into Opportunities

While disputes and legal challenges can be daunting, they're also opportunities to learn and strengthen your processes. By staying informed, keeping detailed records, and seeking professional help when needed, you can avoid a world of pain. If you're talking to someone like me, it means (or it should mean) that there are no other options to try. Not that I'm not happy to talk to you...

Try to focus on resolution of the issue as opposed to proving who is right. If you're right, better to know it and never do business with that person or company again. A win might look like cutting someone out of your life for good, no matter what it cost. A win might just be peace of mind to let you move on with your life.

Alternatively, you can spend the next two years and tens of thousands of dollars to land, maybe, at the same result. I'll always have a job because some people just need to prove they're right. No worries. Either way, I'm looking at beach houses in Lake Tahoe right now.

In the next chapter, we'll discuss how to grow and

scale your content creation business while staying legally compliant.

CHAPTER 13

Building a Legally Resilient Creator Business

The Foundation of Legal Resilience

Building a successful creator business isn't just about producing great content; it's about establishing a solid legal foundation that protects your work, ensures financial stability, and minimizes risks. I'll try to offer a few tips to guide you through the key components of creating a legally resilient business, from structuring your company to implementing (and following) best practices.

No guarantee these will produce long-term success. There are no guarantees in the law or in life. That said, every successful company does pay attention to these things. So, they're worth a read, at least. As you'll see, everything pretty much comes down to common sense. After that, it's mostly tax considerations and risk avoidance (like shielding your personal assets from suits).

1. Choosing the Right Business Structure

The first step in building a resilient business is deciding on the most appropriate legal structure. Each option has its benefits and drawbacks, depending on your goals and level of risk tolerance. I'm not going to dive into a long lecture on these here, but you can find a lot of good information on the internet about the differences.

- **Sole Proprietorship**

  - Simple and inexpensive to set up.

  - Direct control over the business.

  - *Drawback*: No legal separation between you and your business, meaning personal assets are at risk.

- **Limited Liability Company (LLC)**

  - Protects personal assets from business liabilities (especially lawsuits).

  - Flexible management structure.

      o   Not taxed at the entity level (i.e., no double taxation – see corporation description below).

- **Corporation (S-Corp or C-Corp)**

      o   More complex and expensive to set up but each offers significant liability benefits (shields your personal assets from business debts).

      o   Slight differences between S-Corp and C-Corp in terms of how things are taxed. Biggest downside to C-Corp is that the company pays tax on its income, and then you pay tax on the money you get from the company (i.e., double taxation).

      o   Investors generally prefer this structure.

      o   Generally, more formal management processes (e.g., board meetings, minutes, etc.)

- **Partnership**

      o   Designed for 2 or more individuals working together, sharing profits and losses.

      o   Requires a partnership agreement to define roles, profit-sharing, and dispute resolution.

- o Less protections from business debts and suits.

- o No double taxation.

Tip: Consult a lawyer or accountant to choose the structure that best aligns with your business goals. 9 times out of 10, the answer is going to be: "Form an LLC."

---

2. Registering Your Business and Meeting Local Requirements

Once you've chosen a structure, you'll need to:

- **Register Your Business Name**

  - o Check trademark databases to ensure the name is unique and not already in use.

  - o *Example*: A creator who wants to brand their merchandise should secure the trademark for their business name and logo.

- **Obtain Necessary Licenses and Permits**

  - o Requirements vary by location and industry.

  - o *Example*: Some jurisdictions require sales tax permits for selling physical products.

- **Set Up a Business Bank Account**

  - Keep personal and business finances separate to simplify accounting and protect your assets.

  - Remember, "co-mingling" of funds can eliminate the protections you get from LLCs and corporations.

---

3. Building a Strong Legal Framework

A resilient creator business requires clear policies and agreements to protect your interests. Find a good lawyer, tell them you are just starting out and need help that won't break the bank, and see what they can do for you. Most are helpful to new and small businesses. Key documents include:

- **Contracts**

  - Always use written contracts for collaborations, sponsorships, and employee agreements.

- **Terms of Service and Privacy Policies**

  - If you operate a website, these documents clarify how users can interact with your platform and how their data will be handled.

You can find good templates on any number of websites. Pick some company that is similar to yours and start with theirs. If it's too complicated, consult a lawyer.

- **Copyright and Trademark Protections**

  o Register copyrights for your content and trademarks for your brand to strengthen your legal standing. Go spend some time at the USPTO site. It's super-helpful.

Pro Tip: Periodically review and update your contracts and policies to reflect changes in your business or the law. Your lawyer can help you with this.

---

4. Managing Risks and Liability

Minimizing risks is essential for protecting your business and reputation. Here are some ways:

- **Liability Insurance**

  o Generally covers legal fees and damages in case of lawsuits.

  o Insurance isn't free and often it's not cheap, so be forewarned.

- *Example*: Errors and omissions (E&O) insurance can protect creators from claims of negligence or incomplete work.

- **Platform Compliance**

  ○ Adhere to the terms of service (TOS) of platforms where you publish content to avoid account suspensions or bans.

- **Content Review Processes**

  ○ Implement a system to ensure all content complies with copyright, trademark, and defamation laws. Go back and look at your content periodically. Make sure the sand hasn't shifted under your feet since you first created it.

---

5. Diversifying Revenue Streams

A legally resilient business isn't overly reliant on a single income source. Diversify your revenue streams to reduce financial vulnerability:

- **Merchandising**

  ○ Create and sell branded products, ensuring designs are unique and not infringing on others' IP.

- **Courses and E-books**

  o Develop educational materials to share your expertise. For example, you can write a book like this. Oh, who are we kidding? You're not going to write a book.

- **Sponsored Content**

  o Collaborate with brands, using clear contracts to define terms.

  o Cast your net a bit wider when it comes to sponsors. Some have other products, other relationships, and a generally broader interest than you might think.

- **Subscription Models**

  o Use platforms like Patreon to generate recurring income. For some of you, Only Fans may be an option, as well. No judgment, but know what you're getting yourself into (mentally more so than legally). The internet never forgets. Ever. Just ask Paris Hilton.

---

6. Staying Compliant with Tax Laws

Taxes are a critical aspect of running a business. Stay compliant by:

- **Tracking Income and Expenses**

  - Use accounting software to monitor revenue and categorize expenses. Keep receipts, invoices, and bank statements.

- **Paying Quarterly Taxes**

  - Self-employed creators are often required to pay estimated taxes quarterly to avoid penalties. Talk to an accountant about this.

- **Claiming Deductions**

  - Common deductions include equipment, software subscriptions, and home office expenses – also music licenses, camera equipment, paper, business meals, and commuting expenses. If you're a make-up influencer, guess what? All that goop is deductible, too.

Pro Tip: Don't spend your time trying to learn the tax code. Life is too short for that. Hire an accountant or tax professional familiar with creator businesses to optimize your tax strategy. They are more reasonable than you think, cost-wise.

7. Building Your Professional Network

No creator is an island. Surround yourself with professionals who can help your business thrive:

- **Legal Advisors**

  - Consult lawyers for contracts, IP registration, and dispute resolution.

  - Lawyers have networks of clients and connections with other professionals (e.g., bankers, brokers, investors, accountants, etc.). Leverage their network to build yours.

- **Financial Advisors**

  - Work with accountants or bookkeepers to manage finances and plan for growth.

- **Mentors and Peers**

  - Join creator communities to share insights and learn from others' experiences.

8. Adapting to Change and Planning for Growth

The digital landscape evolves rapidly, so adaptability is crucial. There are some things, though, that can help you future-proof your business:

- **Stay Informed**

  - Keep up with changes in platform policies, tax laws, and IP regulations.

- **Invest in Continuing Education**

  - Take courses or attend workshops on business management, marketing, and legal compliance.

- **Develop a Succession Plan**

  - If your business grows significantly, consider how it will operate without your direct involvement.

Building for Longevity

Creating a legally resilient creator business is about more than avoiding pitfalls—it's about setting yourself up for sustained success. By establishing strong legal and financial foundations, diversifying your income, and staying adaptable, you will be way ahead of the curve.

There's an old saying that you should internalize:

"An ounce of prevention is worth more than a pound of cure." Taking a little bit of time to consider the broader legal and tax implications will be dull and painful, I realize.

But, it is way less painful than the colonoscopy you're going to experience if you get audited by the IRS or get sued in court. When you're selling your collection of raccoon statuettes and vintage handbags to pay a judgment against you, the wisdom will become abundantly clear.

Final Thoughts

Your Creative Empire Awaits

Well, look at you! You've made it to the end of this book, which means you're officially a legal nerd now—or at least way more prepared to handle the wild, law-infused world of content creation than when you started. Pat yourself on the back, take a victory lap around your workspace, kiss your dog, and maybe even treat yourself to some overpriced coffee. You've definitely earned it.

What's Next?

Now that you're equipped with all this shiny new knowledge, it's time to put it to work. Whether you're gearing up to launch your first podcast, trademark your brand, or negotiate a killer sponsorship deal, remember this: there are people dumber than you, uglier than you, and shorter than you who have successfully done it. Hey, why not you?

Seriously. The legal stuff might feel daunting at times, but you've got the tools, the tips, and the mindset to tackle it like the creative powerhouse you are.

A Few Final Nuggets of Wisdom

Before we part ways, here are a few final pieces of advice to keep you grounded, protected, and thriving:

1. **Stay Curious:** The internet evolves faster than a meme cycle, and so do the rules. Keep learning, stay updated, and don't be afraid to ask questions.

2. **Play Nice (but Smart):** Collaboration is awesome, but always protect yourself with clear agreements and a healthy dose of trust-but-verify. The world is not a nice place, and jealous jackasses are around every corner – they want you to fail. Learn to spot them, go around them, or, if you need to, go right through them.

3. **Don't Skip the Boring Stuff:** Yes, contracts are tedious, and yes, reading platform policies might make you want to scream into your facial mask thingy. But these details matter. They're what is going to keep you ahead of the losers and jerks.

4. **Laugh When You Can:** If there's one thing this book has tried to teach you, it's that a sense of humor makes the legal journey a lot less painful.

Life's too short to spend on depositions, document review, and court appearances. Get out there and make it happen. If you get a "cease-and-desist" letter, you now know how to push back. With confidence and professionalism. And, a smile.

If it comes to it, you'll hire a good lawyer and go after that bastard who wrote it – and stick that letter up so far up his worthless ass, he'll read it every time he brushes his teeth. Legally speaking, of course.

It's Your Story to Tell

At the end of the day, this whole crazy ride is about your creativity. The legal stuff? It's just the scaffolding that lets you build higher, dream bigger, and share your work with the world. So don't let it intimidate you—embrace it, own it, and use it to your advantage. I have complete faith in you.

And Finally, a Big Thank You

Thank you for sticking with me through this adventure. Writing this book was my way of helping creators like you not just survive, but thrive in the digital age. If it's helped you dodge a lawsuit, secure your rights, or just feel a little more confident about this whole legal thing, then mission accomplished.

Now, go forth and conquer. Create fearlessly, protect fiercely, and start handing out some defiant "L's" to the rest of the competition.

Cheers to your success,

Jeff

If you ever need to get a hold of me, you can find me at:

www.TitanLawNY.com

Sample Confidentiality Agreement

[Just fill in the blanks - also available at <u>www.titanlawny.com</u>]

CONFIDENTIALITY AGREEMENT

This confidentiality agreement (the "**Agreement**") is entered into between _____ [PARTY #1], a [STATE] [BUSINESS TYPE OR INDIVIDUAL] with an address at _____ [ADDRESS] ("**Party #1**") and _____ [PARTY #2], a [STATE] [BUSINESS TYPE OR INDIVIDUAL] with an address at _____ [ADDRESS] ("**Party #2**") with effect as of and from the last signature below ("**Effective Date**").

WHEREAS Party #1 and Party #2 (collectively, the "**Parties**" and each, a "**Party**") wish to explore the possibility of a transaction between the Parties (the "**Purpose**");

WHEREAS one Party ("**Recipient**") or any one or more directors, officers, managers, employees, agents, representatives, consultants and third party contractors ("**Representatives**") of Recipient, may receive or access Confidential Information (as defined below) regarding the other Party ("**Discloser**"), which information is a valuable asset of Discloser;

NOW THEREFORE, the Parties agree as follows:

1.   **Definition**. For purposes of this Agreement, "**Confidential Information**" means information related to any of Discloser (or its subsidiaries) or its business, assets or activities, whenever disclosed, which is not generally known and is or may be used or useful in the conduct of such entity's business, or which confers or tends to confer a competitive advantage over anyone who does not possess the information; in each case, whether disclosed or made available in written, electronic, oral or other form or medium; whether or not marked, designated or otherwise identified as confidential; and whether proprietary of Discloser or licensed by it from a third party.

Confidential Information includes, but shall in no case be limited to (i) any of the technologies, products, business strategies, assets, marketing, businesses, documents, technical, financial, software and hardware, statistical information, customers, investors, pricing, personnel, know-how, trade secrets, intellectual properties, inventions, ideas, methods and discoveries, unpublished patent applications, and/or any other intellectual property, (ii) any information, plans, lists or strategies relating to any one or more of the foregoing subject matters, whether in whole or in part and whether or not related to the Discloser or its affiliates, (iii) any third party confidential information included with, or incorporated in, any information provided by Discloser to any of Recipient or its Representatives, (iv) any memoranda, notices, analyses, compilations, or other writings or records based on the Confidential Information ("*Notes*"), (v) the terms or existence of any negotiation or agreement between the Parties or status of related discussions, or (vi) the equity holders, officers, owners, control or affiliates of the Discloser or its affiliates. However, "Confidential Information" does not include information which (w) was or becomes generally available to the public other than as a result of disclosure by any of Recipient or its Representatives, (x) at the time of disclosure is, or thereafter becomes, available to the Recipient on a non-confidential basis from a third-party source, provided that such third party is not and was not prohibited from disclosing such Confidential Information to Recipient by a legal, fiduciary or contractual obligation to the Disclosing Party, or (y) was known by or in the possession of Recipient or its Representatives, as established by documentary evidence, prior to being disclosed by or on behalf of Discloser pursuant to this Agreement.

2.   **Non-Disclosure**. Recipient agrees that it will, following the disclosure of Confidential Information, keep such Confidential Information strictly confidential and not disclose it to any person for a term of three (3) years following such disclosure; *provided, however*, that any of such information may be disclosed to its Representatives who (i) need to know the Confidential Information to assist Recipient, or act on its behalf, for the Purpose, (ii) are informed by Recipient of the confidential nature of the Confidential Information, and (iii) are subject, at the time of such disclosure, to confidentiality duties and obligations to Recipient that are no less restrictive than those set forth in this Agreement (provided, in the case of any Representative who is not an employee of Recipient, that such duties and obligations must

also be agreed upon in a signed writing). Recipient shall safeguard the Confidential Information from unauthorized use, access or disclosure, and shall use no less than a commercially reasonable degree of care to this end. Recipient will be responsible for any breach of this Agreement caused by any of its Representatives.

3.   **Non-Use**. Recipient shall use the Confidential Information only for the Purpose, but shall in no case use, or cause or permit to be used, the Confidential Information for any purpose detrimental to Discloser (including, without limitation, to reverse engineer, disassemble, decompile or design around Discloser's proprietary services, products and/or confidential intellectual property).

4.   **Additional Obligations**. Except as required by law, neither Party shall disclose to any person, nor cause or permit to be disclosed to any person: (i) that the Confidential Information has been made available to it or its Representatives; (ii) that discussions or negotiations may be, or are, underway between the Parties regarding the Confidential Information or the Purpose; or (iii) any terms, conditions or other arrangements that are being discussed or negotiated in relation to the Confidential Information or the Purpose.

5.   **Required Disclosure**. In the event that Recipient or any of its Representatives is or are requested or required (by oral questions, interrogatories, requests for information or documents, subpoena, civil investigative demand, or similar process) to disclose any Confidential Information, Recipient will give Discloser prompt written notice of any such request or requirement and any documents requested thereby so that Discloser may seek an appropriate protective order. If, in the absence of a protective order, Recipient or any of its Representatives is nonetheless, in the written opinion of Recipient's outside counsel, compelled to disclose any Confidential Information or else stand liable for contempt or suffer other censure or penalty, Recipient or they may disclose such information without liability hereunder; *provided, however*, that Recipient shall give Discloser written notice of the information to be disclosed as far in advance of its disclosure as is practicable, and Recipient/they shall use its/their best efforts, if requested by Discloser and at the expense of Discloser, to obtain an order or other reliable assurance that confidential treatment will be accorded to such information.

6.   **Return or Destruction**.   Upon Discloser's request,

Recipient will promptly return to Discloser or destroy all copies of the Confidential Information and will destroy all Notes prepared by Recipient or any of its Representatives.

7.   **No Transfer**. The Confidential Information of Discloser shall remain the sole and exclusive property of Discloser, and Discloser retains its entire right, title and interest in and to all of its assets and intellectual property (collectively, "**Discloser Rights**"). Neither the present Agreement nor any disclosure of Confidential Information hereunder shall be construed as an assignment, grant, option, license or other transfer of any Discloser Rights whatsoever to Recipient or any of its Representatives.

8.   **No Other Obligation**. The Parties agree that this Agreement does not create an obligation for them to enter into any business or contractual relationship, investment or transaction or to negotiate any terms related thereto. Either Party may at any time, at its sole discretion with or without cause, terminate discussions and negotiations with the other Party, in connection with the Purpose or otherwise.

9.   **Injunctive Relief and Fees**. Each Party agrees that money damages would not be sufficient remedy for any breach of this Agreement by Recipient or any of its Representatives, and that in addition to all other remedies which may be available, Discloser shall be entitled to specific performance and injunctive or other equitable relief as a remedy for any such breach, and Recipient further agrees to waive, and to use its best efforts to cause its directors, managers, officers, members, employees, or agents to waive, any requirement for the securing or posting of any bond in connection with such remedy. In the event a Party institutes a legal action to enforce the provisions hereof and prevails in such legal action, the other Party shall reimburse such Party for all costs and expenses of such legal action and the reasonable fees and disbursements (including attorneys' fees and disbursements) incurred by the latter Party in connection therewith.

10.   **Applicable Law and Forum**. **EACH PARTY HEREBY (a) IRREVOCABLY SUBMITS TO THE NON-EXCLUSIVE JURISDICTION OF ANY _____[STATE] COURT OR FEDERAL COURT SITTING FOR THE COUNTY OF _____ [COUNTY], _____ [STATE] IN ANY ACTION OR PROCEEDING ARISING**

OUT OF OR RELATING TO THIS AGREEMENT AND IRREVOCABLY AGREES THAT ALL CLAIMS IN RESPECT OF SUCH ACTION OR PROCEEDING MAY BE HEARD AND DETERMINED IN SUCH COURT, (b) AGREES THAT COPIES OF THE SUMMONS AND COMPLAINT AND ANY OTHER PROCESS MAY BE SERVED IN ANY SUCH ACTION OR PROCEEDING BY MAILING, REGISTERED OR CERTIFIED MAIL, OR DELIVERING A COPY THEREOF TO SUCH PARTY AT THE ADDRESS PROVIDED HEREIN OR SUCH OTHER ADDRESS AS SUCH PARTY PROVIDES IN WRITING TO THE OTHER PARTY, AND (c) AGREES THAT A JUDGMENT IN ANY SUCH ACTION OR PROCEEDING SHALL BE CONCLUSIVE AND MAY BE ENFORCED IN OTHER JURISDICTIONS BY SUIT ON THE JUDGMENT OR IN ANY OTHER MANNER PROVIDED BY LAW. Nothing herein contained shall affect the right of a Party to serve legal process in any other manner permitted by law or to bring any action or proceeding against the other Party or its property in the courts of other jurisdictions. Each Party hereby irrevocably waives any objection which it may now or hereafter have to the laying of venue of any suit, action or proceeding arising out of or relating to this Agreement in any state or federal court sitting for the County of _____ [COUNTY], _____ [STATE], and hereby further irrevocably waives any claim that any such suit, action, or proceeding brought in any such court has been brought in an inconvenient forum. Each Party hereby agrees to waive trial by jury in any such suit, action, or proceeding. This Agreement shall be governed by and construed in accordance with the laws of the State of _____ [STATE], without giving effect to its conflict of laws principles or rules.

11.		**Notices**. All notices and other communications to be given under or by reason of this Agreement shall be in writing and shall be deemed to have been duly given to a Party: (i) on the day upon which (or, if not a business day, on the first business day following the date upon which) such notice or communication is sent to such Party at the e-mail address of such Party which appears under its signature line below; or (ii) three (3) days after being sent by registered or certified mail to the address of such Party indicated on the first page of this Agreement. Any Party may change its notice address hereunder by giving prior written notice to the other Party of any such change in the manner provided in this paragraph.

12.		**General**. This Agreement contains the entire agreement between the Parties with respect to its subject matter, and supersedes all prior and contemporaneous understandings, agreements, representations and warranties, whether written or oral, with respect to such subject matter. If any term or provision of this Agreement is invalid, illegal or unenforceable in any jurisdiction, such invalidity, illegality or unenforceability shall not affect any other term or provision of this Agreement or invalidate or render unenforceable such term or provision in any other jurisdiction. No waiver by any Party of any of the provisions hereof shall be effective unless explicitly set forth in writing and signed by the Party expressly waiving such provision. No waiver by any Party shall operate or be construed as a waiver in respect of any failure, breach or default not expressly identified by such written waiver. No failure to exercise, or delay in exercising, any right, remedy, power or privilege arising from this Agreement shall operate or be construed as a waiver thereof; nor shall any single or partial exercise of any right, remedy, power or privilege hereunder preclude any other or further exercise thereof or the exercise of any other right, remedy, power or privilege. This Agreement may not be amended or supplemented unless it is explicitly set forth in writing and signed by all the Parties. Neither Party may assign any of its rights or delegate any of its obligations hereunder without the prior written consent of the other Party. Any purported assignment or delegation in violation of this Section shall be null and void. No assignment or delegation shall relieve the assigning or delegating Party of any of its obligations hereunder. This Agreement is for the sole benefit of the parties hereto and their respective successors and permitted assigns. This Agreement may be signed in counterparts, each of which shall be deemed an original, but all of which together shall be deemed to be one and the same agreement. A signed copy of this Agreement delivered by facsimile, e-mail or other means of electronic transmission shall be deemed to have the same legal effect as delivery of an original signed copy of this Agreement. This Agreement shall be of no force or effect against either party unless executed and delivered by both parties.

(signature page follows)

WILSON

IN WITNESS WHEREOF, the Parties have executed this Agreement on the Effective Date:

PARTY #1

Signature:_____

Name: _____

Title: _____

E-mail: _____

Date: _____

PARTY #2

Signature:_____

Name: _____

Title: _____

E-mail: _____

Date: _____

[end of document]

About the Author

Jeffrey L. Wilson, founder and partner at Titan Law PLLC, is a trusted legal advisor known for his meticulous preparation, strategic insight, judgment, and unwavering commitment to his clients. With 20 years of practice in New York City, Jeff has represented public and private clients in complex commercial litigation, guiding them through contract disputes, business conflicts, trials, and appeals in both New York state and federal courts.

Municipalities, government agencies, and corporations have valued Jeff's deep understanding of their unique operational and legal challenges, particularly in finance, insurance, data security, and media. His extensive experience working with New York City and State agencies has equipped him to navigate their distinct mandates and streamline interactions, minimizing administrative delays and fostering effective partnerships.

As a skilled litigator and strategist, Jeff advises clients on deal evaluations, disputes, high-stakes negotiations, risk management, and litigation avoidance. His counsel extends to regulatory matters, company formation, operating agreements, and strategic partnerships. Jeff's ability to identify key contractual vulnerabilities and preserve deal viability has earned him the trust of top executives navigating complex business landscapes.

Jeff has successfully negotiated agreements against high-powered legal teams from industry leaders such as Apple, Ford, Bank of America, NBC Universal, Visa, and The Walt Disney Company.

Jeff earned his J.D. from St. John's University School of Law and a B.B.A. in Finance and Business Economics from the University of Notre Dame. He has received numerous awards, certifications, and honoraria recognizing his client-first approach.

Also, he wears bow ties.

Acknowledgments

I would like to give a special shout out to all of the weird, silly, serious, goofy, racy, shameless, cocky, ridiculous, obtuse, funny, creative, blinky, nervous, jerky, obese, thirsty, morose, unhinged, filter-happy, intimidating, anxious, innovative, engaging, bald, sweaty, authentic, ambitious, strategic, passionate, motivated, versatile, original, quirky, clumsy, insighttful, analytical, beautiful, skinny, wacky, clout-chasing, thoughtful, inspiring, relatable, entertaining, hilarious, visionary, inventive, approachable, dumb-founding, meme-tastic, adaptable, sticky, intuitive, reliable, flaky, perceptive, proactive, resilient, fearless, curious, intelligent, snarky, kooky, dedicated, independent, influential, relentless, trendy, artistic, witty, rude, imaginative, determined, cheeky, persistent, focused, spontaneous, compelling, fast-paced, funky, libertarian, bonkers, giggly, corny, caffeinated, bubbly, charismatic, groundbreaking, balanced, playful, risk-taking, fun-loving, smelly, knowledgeable, nutty, supportive, experimental, informed, driven, assertive, inspirational, high-energy, honest, poised, outlandish, personable, trendsetting, self-reliant, revolutionary, and bold people who create content.

I couldn't have done any of this without you crazy kids.

CHECK OUT OUR BLOG FOR RECENT ARTICLES ABOUT THE NEWEST LEGAL ISSUES FACING CONTENT CREATORS AND DIGITAL INFLUENCERS

www.TitanLawNY.com

TECHNOLOGY MOVES FAST.

SO SHOULD YOUR LAW FIRM.

TITAN LAW

A SMARTER LAW FIRM FOR A FASTER WORLD.

FOLLOW US ON: